THE FUTURE OF LIBERAL DEMOCRACY

THE FUTURE OF LIBERAL DEMOCRACY

THOMAS JEFFERSON AND THE CONTEMPORARY WORLD

Edited by
Robert Fatton Jr.
and
R.K. Ramazani

FUTURE OF LIBERAL DEMOCRACY

First published 2004 by
PALGRAVE MACMILLAN™
175 Fifth Avenue, New York, N.Y. 10010 and
Houndmills, Basingstoke, Hampshire, England RG21 6XS.
Companies and representatives throughout the world.

PALGRAVE MACMILLAN is the global academic imprint of the Palgrave Macmillan division of St. Martin's Press, LLC and of Palgrave Macmillan Ltd. Macmillan® is a registered trademark in the United States, United Kingdom and other countries. Palgrave is a registered trademark in the European Union and other countries.

ISBN 1–4039–6564–1 hardcover
ISBN 1–4039–6565–X paperback

Library of Congress Cataloging-in-Publication Data

Future of liberal democracy : Thomas Jefferson and the contemporary world / edited by R. K . Ramazani and Robert Fatton, Jr.
   p. cm.
Includes bibliographical references and index.
ISBN 1–4039–6564–1 (cl.) — ISBN 1–4039–6565–X (pbk.)
   1. Human rights—Case studies—Congresses. 2. Democracy—Case studies—Congresses. 3. Jefferson, Thomas, 1743–1826—Political and social views—Congresses. I. Ramazani, Rouhollah K., 1928– II. Fatton, Robert.

JC571.F88 2004
323—dc22   2003065828

A catalogue record for this book is available from the British Library.

Design by Newgen Imaging Systems (P) Ltd., Chennai, India.

First edition: October 2004
10 9 8 7 6 5 4 3 2 1

Printed in the United States of America.

Contents

LIST OF CONTRIBUTORS

Akbar Ahmed, Former High Commissioner for Pakistan in the United Kingdom and Ireland, he is currently Ibn Khaldun Chair and Professor of Islamic Studies, American University and has held appointments at Cambridge, Harvard, and Princeton Universities. He has published extensively and produced documentaries on the Islamic world.

Yasushi Akashi, Former Under-Secretary General of the UN, chief of UNPOFOR in Bosnia and Herzegovina in 1993, and former Director of the Hiroshima Peace Institute, is currently Chair of the Japan Center for Preventive Diplomacy, committed to preventing war and conflict expansion.

T.H. Breen, William Smith Mason Professor of American History at Northwestern University and Harmsworth Professor of American History at Oxford University in 2000–2001, has written extensively on British American colonial society and the American Revolution.

Robert Fatton Jr., Julia A. Cooper Professor of Politics and Chair, Department of Politics, University of Virginia, is author of: *Black Consciousness in South Africa* (1986); *The Making of a Democracy: Senegal's Passive Revolution, 1975–1985* (1987); *Predatory Rule: State and Civil Society in Africa* (1992); and *Haiti's Predatory Republic* (2002).

Stanley Hoffmann, Paul and Catherine Buttenwieser University Professor at Harvard University, teaches International relations, their ethics and politics, U.S.–European relations, and French political and intellectual history. His most recent book is *World Disorders* (Roman & Littlefield, 1990).

James Horn, Director of the John D. Rockefeller, Jr. Library at the Colonial Williamsburg Foundation, was formerly Saunders Director of the International Center for Jefferson Studies at Monticello, in which role he organized the conference leading to this volume.

A.E. Dick Howard, White Burkett Miller Professor of Law and Public Affairs at the University of Virginia, was the principal draftsman of Virginia's Constitution, has consulted widely with constitution-makers in other countries, especially in Central and Eastern Europe, and is a prolific author.

Iain McLean, Professor of Politics at Oxford University and Fellow of Nuffield College, has held visiting appointments at Washington and Lee, Stanford, and Yale, and has published on Jefferson, Madison, Condorcet, and the interaction of the American and French scientific Enlightenments.

John M. Owen, Associate Professor of Politics in the Woodrow Wilson Department of Politics, University of Virginia, is the author of *Liberal Peace, Liberal War: American Politics and International Security*.

Jack N. Rakove, Coe Professor of History and American Studies, Stanford University, winner of the Pulitzer Prize for *Original Meanings: Politics and Ideas in the Making of the Constitution*, is a specialist on political rights and the U.S. constitution.

Nesta Ramazani, Independent writer and lecturer, has published numerous articles on women's rights in the Middle East. She has lectured on the subject on behalf of the United States Information Agency in Israel, Germany, Pakistan and Turkey.

R.K. Ramazani, Edward R. Stettinius Professor Emeritus of Politics at the University of Virginia, is a recipient of the Thomas Jefferson Award and has served on numerous governing boards, including those of the Middle East Institute and the American–Iranian Council. He is an internationally recognized authority on Iran and the Persian Gulf and has authored numerous books and over a hundred articles and book chapters.

Alan Ryan, Warden and Professor of Politics at New College, Oxford, has written on the philosophical premises of liberalism and British and American rights.

Soli Sorabjee, Attorney General of India, is the leading human rights, lawyer in India and has played an important role in many international judicial colloquia.

Mark Tamthai, Professor of Philosophy, and Director of the Humanities for Society Center, Chulalongkorn University, Bangkok, and Chairman of the National Security Council's sub-Committee on Strategic Non-Violence and Foreign Policy. His recent publications are on the democratic development of constitutional monarchies.

Jose Thompson, Director of the Center for Electoral Assistance and Promotion at the Inter-American Institute of Human Rights, he has published extensively on human rights in Latin America.

Verma, J.S., Chairperson of the National Human Rights Commission of India, a justice of the Supreme Court, and former Chief Justice of India. A collection of his speeches has been published as a book entitled *New Dimensions of Justice*.

Brantly Womak, Professor of Comparative Government and International Relations in the Woodrow Wilson Department of Politics, University of Virginia, is the author of many books and articles on Chinese and Asian politics.

PREFACE

Of the transcendent principles that gave rise to the great wave of revolutionary fervor, which swept across Europe and the Americas in the late eighteenth and early nineteenth centuries, none figured so prominently as the issue of rights. For Thomas Jefferson, rights, collective and individual, were the very basis of political philosophy, the foundation of his thinking about civil society and the relationship between the individual and the state. The extended experiment in republicanism that took place after the American Revolution, involving the replacement of a sovereign monarch with a sovereign people capable of participating in representative democracy, necessarily entailed an intense scrutiny of the political rights of the people, both in the exercise of their responsibilities as citizens of the new republic and to establish a constitutional bulwark that would guard against the possibility of an irreversible erosion of the citizens' rights in the future. Jefferson was not alone of course in this undertaking—all of the Founding Fathers as well as some leading European intellectuals were involved—but Jefferson went a great deal further than most in extending political rights into social and economic spheres. Proposals such as the provision of free land for poor householders, a state-sponsored education system, changing inheritance laws to undermine hereditary wealth, garnered only limited support from his fellow lawmakers in his native state of Virginia, yet the breadth of his thinking would be echoed in the reformist impulse of the French Revolution, subsequent revolutions of 1830 and 1848, and, more broadly, in the social welfare programs of the late nineteenth and twentieth centuries.

If, as Louis Henkin has suggested, "human rights is the idea of our time," here as in so much else, Jefferson was ahead of his times. The Declaration of Independence is surely a foundation of our modern ideas of human rights, with its very language echoed in the Universal Declaration of Human Rights. Today the issue of rights lies at the core of pro-democracy movements sweeping across every major region of the world including Asia, South America, the Middle East, and Africa. The process of decolonization after World War II, the end of the Cold War and disintegration of the Soviet Union, combined with landmark advancements in science and technology and the explosion of information technology have globalized peoples' consciousness of the issue of human rights to an unprecedented degree.

Jefferson could not have possibly foreseen the efflorescence of human rights legislation and organizations of the last half-century but he fervently hoped that the benefits of enlightened government, enjoyed by Americans in his lifetime, would one day be enjoyed by all. This would be the means of eventually eradicating the gross inequalities, ignorance, and many of the social and economic ills that bedeviled traditional societies. "That form [republican government] which we have substituted," he wrote in his last letter shortly before his death, "restores the free right to the unbounded exercise of reason and freedom of opinion. All eyes are opened, or opening, to the rights of man.... These are grounds of hope for others."

This volume evolved from a conference sponsored by the International Center for Jefferson Studies at Monticello, in collaboration with the University of Virginia, on "Thomas Jefferson, Rights, and the Contemporary World," held at the Bellagio Conference and Study Center in June 2002. The International Center is committed to advancing Jefferson studies worldwide, and in pursuit of that mission has recently initiated a series of international conferences devoted to exploring significant contemporary issues that in some measure reflect Jefferson's own ideas and ideals, with the aim of creating a forum for wide-ranging discussion between distinguished public leaders, practitioners, and scholars. Accordingly, the chapters in this volume take a variety of forms, including lengthy analyses by academicians as well as shorter pieces by practitioners and prominent officials involved in human rights legislation and policy-making, together encouraging a vital dialogue between theory and practice.

A volume such as this could not have been completed without the support of many colleagues and institutions. The inspiration for an international conference series came from the late Mr. Rodman C. Rockefeller, who generously gave his time and support to promote the venture. Anthony, Lord Lester of Herne Hill, QC, President of the International Centre for the Legal Protection of Human Rights and Liberal Democrat Peer in the House of Lords, was unstinting in providing invaluable advice about possible participants, and R.K. Ramazani, Edward R. Stettinius Professor Emeritus of Government, and Foreign Affairs and A.E. Dick Howard, White Burkett Miller Professor of Law and Public Affairs, of the University of Virginia, also gave early encouragement and direction. The International Center is indebted to The Rockefeller Foundation, especially to Susan Garfield and staff of the Bellagio Study and Conference Center for making our stay at the Villa Serbelloni so comfortable. As anyone who has had the good fortune to visit can attest, the superb location and facilities of the Center play an important role in making work and exchange of ideas so productive. I am grateful to the editors of the volume, professors R.K. Ramazani and Robert Fatton together with the authors, for bringing the project to timely completion, to Michael Smith and James Sofka of the University of Virginia, to Laura Gross and Kathryn B. Knisley, former and current administrative assistants at the International Center for Jefferson Studies, and to Ella Pearce and her colleagues at Palgrave Press. Finally, I would like to record my own debt of gratitude to Daniel P. Jordan,

President of the Thomas Jefferson Foundation, for his unflagging commitment to the project. Without his support, and that of the Foundation, neither the conference nor the volume would have been possible.

James Horn
September 2003

INTRODUCTION

Robert Fatton Jr. and
R.K. Ramazani

The essays in this book are the product of a conference entitled "Thomas Jefferson, Rights, and the Contemporary World," held in Bellagio, Italy, in June 2002. The conference was organized by the International Center for Jefferson Studies with the support of the Rockefeller Foundation. The dawn of the new millennium had created for many of the participants a sense that a new American century and empire were in the making, one in which liberal values seemed to be on the verge of triumph. On the other hand, the new millennium also announced a wave of violent terror bent on resisting the global "Pax Americana." The year that followed the conference generated new conflicts and weakened old alliances. Suddenly there were fissures in America's hegemony; the facile certainties of the 1990s generated confusion and prior assumptions about the future world order were called into question.

In the early 1990s, the end of the Cold War and the collapse of Soviet communism precipitated the ascendancy of the "liberal moment" and the euphoria of its advocates. Inexorably, it seemed, Western-type democracies were taking roots throughout the world, displacing outmoded and brutal dictatorships. The human rights revolution was under way and it expressed what many took to be the universal aspirations of Jeffersonian and Madisonian constitutionalism. Not surprisingly, the liberal moment embodied for some the "end of history" and the definitive triumph of the United States.[1] In this vision, America had become the lone superpower whose interests and ideology were ultimately congruent with those of the rest of the world. Thus, America was distinctively unique because it exercised its "hyper-imperialism" benignly, promoting individual rights, multiparty systems, and market economies. These attributes were deemed as universally desirable and applicable to all nations irrespective of geographical location, cultural traditions, or historical legacies. Democratization and human rights were therefore on the global agenda; the United States was merely leading the world to its inevitable destiny.

It is clear, however, that this deterministic vision of democratic transformations, which reached its most messianic articulation in the 1990s, had many vocal critics. First, the idea that American imperialism is benign and well intentioned has certainly not become universal. Irrespective of motivations, there is widespread resentment and opposition to the "hyper-power" of the

United States. The tragic events of September 11, 2001 have come to symbolize the most hostile and violent rejection of America's imperial reach. There are less extreme but no less visible signs of opposition, such as "old Europe's" determined "no" to the U.S. war against Iraq and the massive global manifestations for peace that preparation for the war provoked. Huge majorities see the American empire as embodying the vices and excesses of any empire, and they greet its claims of benignity with skepticism if not outright derision. From abroad, what is "good for America" is not necessarily viewed as good for the rest of the world. The universal sympathy for America that followed September 11 soon evaporated because of Washington's growing unilateralism and predisposition to use military means to combat terrorism. While many American scholars, like John Owen in this volume, still maintain that America's strategic goals have a strong overlap with the spread of democracy in hitherto dictatorial societies, elsewhere few are convinced. Abroad, analysts tend to perceive such claims as cynical justifications for a coercive Pax Americana, or as naïve Panglosian hymns to impossible dreams.

In reality, the process of democratization is much more complicated and difficult than the euphoric optimism of the early 1990s ever admitted. It is not merely that the legacies of authoritarianism tend to impede the liberalization of politics and the consolidation of democracy like a proverbial ball-and-chain, but that democracy and its companion, human rights, are themselves contested concepts. Moreover, the idea that liberal democracy can be "exported" through the imposition of foreign constitutional designs, diplomatic and economic pressures, and/or the compelling force of the bayonet has become increasingly doubtful. Such an "export" of democracy is more likely to generate a backlash against the West, and particularly America, than a consensual embrace of a Kantian peace of free nations. The "export of democracy" simply ignores the fact that specificities of culture and history inevitably shape notions of freedom, rights, and the individual's place in the community. If democracy is to triumph, it has to be rooted in its local milieu, and thus it will take on the multiple faces of a world full of diversity and differences. Finally, the severe poverty besieging a vast portion of humanity and the basic socioeconomic inequities characterizing both the world system and national polities bode poorly for any meaningful pattern of democratization. There are thus powerful and obdurate constraints on the global spread of democracy, liberal or otherwise.

These constraints, however, are not absolute; they simply indicate that genuinely representative forms of governance facilitating the flowering of human rights are difficult to establish and may require a long and arduous journey. This book is an attempt to assess the degree of difficulty in accomplishing the journey. It explores from varied, and at times conflicting, perspectives the theoretical, historical, strategic, and practical aspects of this journey.

In part I of the volume, Iain McLean explores the influence of Thomas Jefferson on the making of three seminal documents in the history of human rights; the American Constitution and Bill of Rights, and the French

Déclaration des Droits de l'Homme et du Citoyen. McLean argues that while Jefferson's critical role in the writing of the first two is well known, it is grossly underestimated in the third. Jefferson was in Paris during the period when the *Déclaration des Droits de l'Homme et du Citoyen* was elaborated and he used his connections with the French intelligentsia to influence its content and spread his republican ideas. According to McLean, however, Jefferson did not "believe that France, only just abandoning feudalism, was ready for a declaration of rights as thoroughgoing as he was pressing on the US ratifying states." This was one of the ironies of Jefferson's "deep contradictions on human rights," for his French hosts "knew that a slaveholder was chiding them for their backwardness in adopting the rights of man—a slaveholder who thought that they were not ready for a declaration of rights as advanced as that of Virginia." It is clear that Jefferson's eagerness to spread the liberal gospel of human rights, albeit contradictory and two-faced, is deeply incrusted with what can be called America's imperial tradition. This should not surprise us, for as Alan Ryan argues, liberalism "is intrinsically imperialist."

Ryan, who contributes the second essay in this collection, resists the intrinsic logic of liberalism and insists that liberals should not succumb to the powerful attraction of the imperialist temptation. He draws this conclusion from "low reasons of practicality" rather than moral principles. Liberals should not apologize for wanting to spread "freedom and enlightenment," in fact they have the right to do so. The problem, however, is that the use of military means to achieve liberal goals simply does not work. Ryan invokes multiple reasons for this failure; chief among them is the extraordinarily difficult attempt to insert liberal ideals into illiberal cultures. He thus recommends "a nonmilitary but morally uninhibited global liberalism, *if it were pursued intelligently and consistently.*"

Ryan's prudence rests mainly on the cultural argument that liberal values are so fundamentally foreign to non-Western dictatorships that their successful implantation in these societies, if at all feasible, is a long-term and exceedingly complex affair. While acknowledging the difficulty of the enterprise, Jack N. Rakove in the third essay of this collection argues that the universal adoption of Jeffersonian values would generate a much-improved moral and political order for everyone in the world. It is not that Jefferson represents the supreme paradigmatic authority on rights and democracy; but rather that in spite of the "embarrassing contradiction between his abstract commitment to equality and his self-indulgent life as the slave-owning master of Monticello," he engineered with James Madison's help an appealing solution to the threat of theocracies.

The solution in Rakove's eyes consists of transforming religious belief into a matter of individual choice. This privatization of the sacred prevents the state from acquiring too much power and from imposing on citizens a single "religious mode of reasoning." The result is limited government, religious diversity and competition in civil society, and the gradual development of a culture of "mutual toleration and respect." There is no doubt in Rakove's mind that universal adoption of such a solution would make the world a "better, safer place." This is so, he argues, because "the basic insight on which the principle of freedom of conscience rests is the recognition that the religious beliefs of *all* faiths are necessarily a matter of inner opinion, and that opinion cannot be coerced or accurately measured or ascertained." To that extent, the Jeffersonian concept of freedom of conscience on which is built the "high wall" of separation of church and state is a fundamental bulwark against religious dogmatism and violence. It protects heretics and shields dissent from the abuses of fundamentalist clerics. Simply put, democracy is impossible without freedom of conscience. The critical question for Rakove, therefore, is whether democracy is an exportable commodity. Rakove is openly ambivalent on this matter; he is hopeful that Jeffersonian principles and ideals are not merely a "manifestation of American exceptionalism."

In his chapter 4, T.H. Breen expresses no such doubts; he argues that the "language of rights" that generated the "Lockean Moment" of America's violent war of independence, has universal appeal. As he puts it: "once a society takes on board the language of rights, it puts on the defensive those people who would seek to exclude others." The point is that "Lockean Moments" can occur in any society in which authoritarianism and inequities exist. Moreover, Breen asserts that the language of rights "may be the best thing on offer to revive a meaningful capacity for solidarity and resistance against global forces that have little use for talk of human rights."

Breen's argument is echoed in Akbar Ahmed's comparison of Thomas Jefferson and Muhammad Ali Jinnah, the respective founding fathers of America and Pakistan. He contends that "Lockean Moments" are not the sole property of the West. In fact, Ahmed posits, "religious freedom and respect for universal education are not exclusive to Western civilization but also to be found in Islamic civilization." He rejects the popular and dominant view that Muslim and Judeo-Christian worlds are inevitably condemned to a "clash of civilization."[2] The example of Jinnah's humanism demonstrates that "Islam is clearly compatible with democracy." Moreover, in Ahmed's words "there can be a genuine meeting of minds between Islam and the West." Thus, a "dialogue of civilizations is not only desirable but also possible."

The possibility of such a dialogue does not, however, mask its difficulties. This is especially the case in developing international mechanisms that might ensure respect for human rights and deter abuses of power. In chapter 6, Justice J.S. Verma points out the necessity of developing "a permanent system of international criminal justice." Such a system would have universal reach to punish individuals for gross criminal behavior. The problem, however, is not merely one of developing a comprehensive and authoritative body of international law, but also of creating a "credible enforcement mechanism." To do so, offenders would have to know that there is no safe haven for them wherever they might be. A good case in point is General Augusto Pinochet, the former dictator of Chile, who was arrested while in England, for committing crimes against humanity. As Justice Verma emphasizes, this case set "a precedent for limiting claims of immunity by former heads of state and thus [opened] the way for their more general prosecution." While the Pinochet case might deter current political leaders from terrorizing their citizens, Justice Verma is concerned that rulers of mighty nations will always be able to escape international punishment even if they are guilty of criminal activities. In Verma's view "there is need for settled norms to prevent [the misuse of international law] by the powerful against the not so powerful, as also for its equal application." Accountability should not be demanded solely of the weak or relatively weak, it has to be required of all. He thus calls for a justice that is truly universal: "An effective mechanism to administer justice by punishing equally all crimes against humanity alone can secure accountability for those crimes. The mechanism has to be equally efficacious against all offenders, irrespective of the power of the State to which they belong."

The spread of a culture of human rights is not merely a matter of uniformly enforcing international law; it is also a complex process of constitutional engineering in environments that can be hostile at times. In the seventh essay in this collection, Soli Sorabjee explains that India's constitutional crafting had domestic roots. Indeed, Lokmanya Tilak, one of the most important figures of the struggle for Indian independence was the major inspiration for the country's Bill of Rights. Like Jefferson, Tilak was convinced that a Bill of Rights was necessary to protect the individual from potential abuses from the state. Moreover, Sorabjee contends that India's Bill of Rights is a living document that has changed to accommodate new circumstances and expand the freedoms and rights of groups that had hitherto been marginalized. India's democracy and constitutionalism are thus rooted in its local milieu, even if they borrow certain foreign norms and principles.

In his examination of rights in the nations of Central and Eastern Europe, Dick Howard stresses the advantages of having domestic legacies compatible with the development of, and respect for democratic laws. As he puts it:

> History helps explain the fact that constitutionalism and the rule of law seem more deeply imbedded in some of the post-communist countries than in others. Those lands that lay within the Habsburg domain fell heir to Austrian traditions of a *Rechtstaat*. The Austro-Hungarian Empire may have been notorious for its bureaucratic tendencies, but it developed a strong tradition favoring the rule of law in its successor states in Central Europe. Peoples under Ottoman rule were not so fortunate.

Howard, nonetheless, emphasizes the difficulties that all post-communist nations face in their transition from dictatorial rule to democratic governance. The communist heritage contributed to the utter underdevelopment of civil societies and engendered strong popular cynicism that undermined citizens' full participation in politics. In their democratic journey, Howard concludes, the countries of Eastern and Central Europe "must travel a rougher road than many in the West might realize." This is not to say that they cannot establish long, lasting liberal democracies and that there are no encouraging signs, but rather that historical legacies constrain political crafting.

Yasushi Akashi, who has personally dealt with acute political crises, particularly in Asia, emphasizes the dilemmas of creating viable democratic structures in post-authoritarian and conflict-ridden countries. As head of the United Nations peacekeeping operation in a devastated Cambodia, he was fully aware that the realities of material scarcity and underdevelopment prevent the emergence of a large middle class and endanger democratic consolidation. His Cambodian experience did not lead him to believe, however, "that development and democracy were incompatible with each other or that human rights and the unity of state could not be harmonized with each other." Akashi is thus prudently optimistic about the possibility of democratic breakthroughs in the Third World.

In chapter 10 Mark Tamthai analyses the contemporary situation in Thailand. He shares Akashi's general optimism but stresses that the "fledgling democracies" of the Third World confront the problems of ensuring human rights in a "hierarchical" environment of deep "structural injustices." To remedy this situation, Tamthai advocates the adoption of a "deliberative democracy" in which citizens engage in deliberations to attempt to elucidate a "common good" acceptable to all. In this sense, democracy is rooted in those local traditions that can enhance a sense of solidarity and good will. It cannot be a copy of an alien model. Not surprisingly, Tamthai suggests that Thailand should appeal to certain principles of its own Buddhist tradition to foster human rights and democracy while it could select some themes from, if not altogether ignore, the Jeffersonian legacy.

In chapter 11 on women rights in Iran, Nesta Ramazani shows, however, that the implementation of certain aspects of local culture and religion can disempower large segments of society. In terms of health, literacy, and life expectancy there are promising signs for gender equality in Iran. The fact remains, however, that the adoption of Islamic law—*Shari' a*—has imposed severe social strictures on women's freedoms. For instance, a woman can marry neither a non-Muslim man nor a foreigner without "special permission." Moreover, women have to confront the possibility that a male guardian may consent to give them in marriage at the young age of 13. There is also the continuing problem of polygyny, which institutionalizes patriarchy even if it is not widespread in the country. Women are thus faced with a series of challenges, but it is clear that they are willing to meet them. As Ramazani puts it:

> [Iranian women] are not simply rebelling against social strictures, but also against the tyrannical silencing of dissent, whether manifested through speech, in print, or in dress. Yearning to have a voice in the governing of their country, large numbers of students, increasingly disenchanted with the slow pace of reforms, have demonstrated, at great personal risk, for a civil society, the rule of law, freedom of speech, and freedom from arbitrary detention, imprisonment, and torture.

Iranian women have a strong sense of realism about their capacity to effect change. They know that the transformation of culture and of personal behavior is a long-term process. While Iranian women are not resigned to their fate and are confident that they will ultimately triumph, Ramazani reports that they are conscious that their full empowerment "will not happen in their lifetime."

In another chapter on Iran, R.K. Ramazani similarly emphasizes the century old democratic aspirations of the Iranian people. He suggests that Islam and Western liberal principles are compatible in the eyes of secular and religious "modernists." He underscores, however, the continuing vitality of Iran's entrenched autocratic culture and the difficulties of defeating it. Ramazani argues that two previous attempts to defeat "the culture of dictatorship" failed in the twentieth century, and that both internal conservative forces and the current policies of the Bush administration seriously endanger an ongoing third. Began in 1997 with the overwhelming electoral victory of Mohammad Khatami, this third democratic attempt is fraught with ambiguity. Ramazani argues that President Kahatami's conception of "Islamic Democracy" is "entangled in . . . contradictory principles" because it accepts the "doctrine of the rule of the religious leader." Thus, it equivocates between the rule of the faqih and popular sovereignty.

In fact, by privileging Islam and downplaying the democratic and pre-Islamic traditions of Iran, "Islamic Democracy" has lost its popular appeal. While conservative religious forces are again politically ascendant, they have been unable to stop a raging debate on the relation between state and religion. According to Ramazani, this debate "has never before been as widespread, as sophisticated and as unrementing as it is today." Thus, in his view the path to "Iranian Democracy" following the two fundamental principles of justice and liberty is still possible. It faces, however, the hostility of domestic theocratic forces and the foreign threat of President Bush's neo-conservative policies.

As Ramazani puts it: "The reality of unprecedented American military presence in Afghanistan, Iraq and Central Asia with thousands of miles of borders with Iran is perceived by Iran as encirclement by an enemy that is the world's sole superpower." This encirclement coupled with Washington's bellicose statements symbolize the Bush administration's utter incapacity to understand that "in the Iranian political culture the quest for democracy has been for more than a century intertwined with preserving Iran's political independence" Instead of implanting democracy in Iran, America's hegemonic and military pretensions can only bring catastrophe for the region.

The problems of American-led exports of democracy affect not only the Middle East; they are quite visible in other areas of the world, such as Haiti for instance. Indeed, as Robert Fatton Jr. makes clear in his essay, the 1994 American military restoration of the elected regime of Jean-Bertrand Aristide failed to uproot the island's authoritarian tendencies even though it enjoyed the widespread popular support of the Haitian masses. This failure is a powerful warning about America's capacity to transform foreign countries; it indicates that in certain situations, internal factors play a more determinant role in the making and unmaking of democracies than international constellations of forces.

In fact, Fatton argues that the primary obstacles to Haiti's democratization are rooted in the country's domestic political economy. He contends that the acute material scarcity besieging the island has contributed to the implantation of *la politique du ventre,* the politics of the belly, whereby different factions of the Haitian political class have traditionally vied with each other to "eat" the limited fruits of power. *La politique du ventre* represents a form of governability based on the acquisition of personal wealth through the conquest of state offices. Given that poverty and destitution have always been the norm, and that private avenues to wealth have always been rare, politics became an entrepreneurial vocation, virtually the sole means of material and social advancement for those not born into wealth and privilege. Controlling the state turned into a zero-sum game, a fight to the death to monopolize the sinecures of political power. Moreover, material constraints have prevented the development of both a productive bourgeoisie and a large working class—the classes whose struggles have historically resulted in liberal democracies. Finally, scarcity has nurtured an authoritarian *habitus* that has facilitated the development of personal rule and presidential monarchisms. Fatton concludes by arguing that countries facing the constraints that have besieged Haiti are more likely to engender what he calls "predatory democracies" than classical liberal democracies.

In chapter 14 on Latin America, José Thompson similarly underscores the "dark side" of democracy. While he acknowledges the democratic explosion of the 1990s and the fundamental role played by nongovernmental organizations (NGOs) in the process, he points out the significant shortcomings of existing forms of political representation and participation. It is true that elections have taken place virtually everywhere in Latin America and that civilian regimes have replaced dictatorships, but gross patterns of economic inequalities and social marginalization persist. Women, indigenous peoples, and Afro-Latin Americans continue to face major obstacles to their full emancipation, they simply "lack a sense of belonging"; the coming of democracy has not dramatically changed their subordinate conditions. Not surprisingly, Thompson argues, "growing disillusionment with democracy is increasingly in evidence. Public opinion polls consistently find lowest ratings for legislative bodies and political parties, without exception, and reveal a dangerous yearning for the easy solutions of authoritarian days." Thus, Thompson reminds us that democracy is more than just elections; it entails effective political participation of all as well as nurturing the full flowering of human rights in their broadest sense.

Many Latin American countries thus face the danger of joining the ranks of the "failed-states" that have proliferated since the collapse of the Soviet Union and the end of the Cold War. The democratic wave of the 1990s and the hopes for global stability that it generated have given way to an increasing sense of global disenchantment. As Stanley Hoffmann explains, the illusions of "a world united against aggression under the leadership of the Big Five Security Council powers" were soon lost with the worldwide recrudescence of ethnic conflicts and old hatreds. In fact, the disintegration of many legally sovereign nations generated a pattern of "humanitarian interventions" that failed to prevent genocides and met with increasing opposition. These failures, however, took place in the traditional structure of the state-system; people were either fighting against new claims of sovereignty or aspiring to create their own nations. Hoffmann contends, however, that things changed dramatically on September 11, 2001. He writes:

> We knew that modern weapons made security behind borders impossible to preserve. We now know that non-state actors, private groups or gangs, armed with weapons as unexpected as hijacked civilian aircraft, could settle almost anywhere and strike deliberately, civilians above all, and spread terror. Terrorism is not new, but it had mainly been an internal phenomenon except when the terrorists were serving a state that wanted to strike far from its borders. Now we can talk about a universal war that knows no borders, which makes the idea of victory perfectly unrealistic.

Hoffmann fears that we may have entered into a global "jungle" where "empire is unlikely [and] peace impossible." He raises serious doubts about the capacity of the Bush administration to guide the world toward a new period that would privilege "the fight against and prevention of cruelty, oppression, fear, misery and injustice." In his view, the unilateralist and imperialist predispositions of Washington bode poorly for a successful taming of the beast unleashed in September 2001.

This is a view that Brantly Womack wholeheartedly supports. In chapter 16, "The United States, Human Rights, and Moral Autonomy in the Post–Cold War world," he argues that American unilateralism is bound to generate resistance and weaken its capacity to lead. Womack writes: "Because the center serves no interest greater than its own, it gives everyone else an incentive to reduce the power and discretion of the center. The center commands, but it does not lead. The rest comply, but they do not cooperate." He argues for a regime of international cooperation that would subordinate immediate imperial gratification to long-term objectives; such a regime would not undermine American interests but establish the basis for America's "sustainable leadership."

Womack contends that large asymmetries of power provoke significant infringements on the moral autonomy of the weak. This is particularly the case for human rights. Not surprisingly, given the extent of its military and economic supremacy and its predisposition to act unilaterally, America has sought to impose on others its own conception of human rights. In doing so, Womack argues, the United States violates the moral autonomy of weaker nations without having to submit to other societies' moral challenges. He advocates a different approach to human rights, an approach that emphasizes cooperation and multilateralism and thus respects more carefully the moral autonomy of each nation. To do otherwise is self-defeating; Washington must understand that it does not enjoy absolute power; it should remember the lessons of Vietnam. Womack claims: "A preponderance of power does not necessarily create the ability to dominate others, and the long-term interests of even a superpower require broad cooperation." Unilateralism is bound to generate "the confounding issue of external imposition," and thus resentment and resistance against America's global reach.

John Owen offers a contrasting perspective in the final essay of this volume, "Human Rights, Peace, and Power." He sees the exercise of America's power coinciding with the expansion of human rights and freedoms. In his view, there is no contradiction between the global spread of liberal democracies and the American empire. Owen contends: "The growth in the number of states that uphold human rights has entailed an expanding zone of international peace and a rise in the influence and power of the United States." This is so, according to Owen, because liberal democracies share a form of politics and philosophical principles that are virtually incompatible with the waging of war against each other. The triumph of liberalism and human rights thus overlaps strongly with American hegemony and a peaceful world order. In Owen's words, "the liberal-democratic club tends to be a zone of American influence... [Liberals] do not worry about the purposes to which America is likely to put its power; or at least, they do not worry enough to favor devoting their own countries' precious resources to counterbalancing U.S. power." America, however, tends to use its diplomatic, economic, and military arsenal to oppose regimes it defines as "evil" or "illiberal." Not surprisingly these are the regimes bent on undermining the United States' primacy. Owen argues that such regimes fear human rights because they are the emblem of America's global reach and imperial predispositions. He thus concludes: "liberal democracy is an instrument not only of peace but of U.S. hegemony."

It is clear that the contributors to this volume articulate conflicting and at times contradictory philosophical approaches to the profound complexities confronting a diverse, dangerous, and unequal world system. This should come as no surprise because even if conceptions of human rights and democracy contain universal truths, they are contested constructs built on different understandings of history.

NOTES

1. Francis Fukuyama, *The End of History and the Last Man* (New York: Free Press, 1992).
2. Samuel Huntington, *The Clash of Civilizations and the Remaking of World Order* (New York: Simon and Schuster, 1996).

PART I

HUMAN RIGHTS, THOMAS JEFFERSON, AND THE LOCKEAN MOMENT

CHAPTER 1

THOMAS JEFFERSON, JOHN ADAMS, AND THE DÉCLARATION DES DROITS DE L'HOMME ET DU CITOYEN

Iain McLean

I am among those who think well of the human character generally. I consider man as formed for society, and endowed by nature with those dispositions which fit him for society. I believe also, with Condorcet . . . that his mind is perfectible to a degree of which we cannot as yet form any conception.

Thomas Jefferson to William Green Munford, June 18, 1799, in Peterson (1984), 1064

All doors of all departments were open to him at all times, to me only formally and at appointed times. In truth, I only held the nail, he drove it.

Thomas Jefferson on his relations with the Marquis de Lafayette in 1789, from speech at banquet in honor of Lafayette, Charlottesville, VA, November 20, 1824, in Malone (1951), 46

INTRODUCTION

Thomas Jefferson lives, as John Adams said on July 4, 1826, a few hours after Thomas Jefferson died and a few hours before John Adams died. Among other things, he lives through his direct influence on constitutional design. In the field of human rights, he influenced both the U.S. Constitution and Bill of Rights (especially the First Amendment), and the French Declaration of the Rights of Man and the Citizen of 1789 (DDHC). The purpose of this chapter is to examine Jefferson's role in the French declaration. It is a role that has been seriously underestimated both by American scholars who do not read French and by French scholars unwilling to admit that their revolution was not homegrown.

The five years that Jefferson spent as American minister in Paris (1784–89) represent an extraordinary conjunction of the French and

American Revolutions. Jefferson arrived in summer 1784, together with
John Adams, to join Benjamin Franklin and form a three-person American
Ministry in Paris. In 1785, Adams went to London, Franklin returned to
Philadephia, and Jefferson remained as sole American minister in Paris until
his departure in September 1789 after witnessing some of the opening scenes
of the French Revolution.

Jefferson's sojourn was equally a confluence of two rivers of the scientific
Enlightenment. From a common fount a century earlier they had diverged,
but reunited in the salons of Mme Helvétius and Sophie de Grouchy,
Marquise de Condorcet.[1] Jefferson and the Marquis de Condorcet met regularly in Paris and admired one another.[2] More generally, it was a time of
fruitful cultural exchange. Where would American architecture have been
but for Jefferson's books and sketches from Europe? Would Americans still
think Madeira was the finest European wine if Jefferson had not introduced
them to Médoc? Where would the Library of Congress and the art and science of bibliography have been, had not Jefferson collected books so eagerly
in Paris and then sold his library, and presented his catalog (rediscovered in
1989), to Congress in 1815?

There is enough human interest in the story of Jefferson in Paris to have
persuaded Ismail Merchant and James Ivory to film it (moderately accurately). The recently widowed Jefferson went to Paris in 1784 with his eld-
est daughter Martha. On hearing in 1785 that his youngest daughter Lucy
had died of "a most unfortunate Hooping Cough," he planned elaborately
for his remaining child Maria (Polly) to join them in the care of his young
slave Sally Hemings (a job she was too immature to do, according to Abigail
Adams[3]). Sally Hemings was his late wife's half sister. DNA (Y-chromosome)
analysis has, however, proved that the child she bore in Paris was not
Jefferson's (Foster 1998). While in Paris, Jefferson fell in love with Maria
Cosway, the flirtatious Anglo-Italian wife of a gay English painter, but in his
Dialogue between my Head and my Heart (1786) addressed to her,
Jefferson's Head suppresses his Heart. There is no evidence that the Heart
ever had its way, then or later. In the French Revolutionary Terror of
1793–94, Maria Cosway retired to a convent to run a school.

For a long time the French historiography of the Revolution was a return
to the barricades. The Revolution was seen through the lens of the author's
position in contemporary French politics. This did not make for good historiography. So, when Jellinek (1902) first suggested that the DDHC was
strongly influenced by the American Revolution and American Revolutionary
ideas, he was denounced as a foreigner who had no right to appropriate the
sacred symbol of *la gloire* (cf. Boutmy 1902). Jellinek was quite right. But
when French scholars have returned to look (however reluctantly) for the
American influence on the DDHC, they have looked in the wrong place.
Ignoring the obvious facts that Jefferson was in Paris, and John Adams either
in Paris or in London, for the whole material time, they have looked for
influences in the American state constitutions and in the reports reaching
France about the drafting and ratification of the U.S. Constitution, while

paying astonishingly little attention to Jefferson's barely concealed under-mining of the court to which he was accredited. For example, not a single contributor to the bicentenary essays on the Declaration in Colliard (1990) cites the Jefferson Papers (Boyd et al. 1950–), in which Thomas Jefferson's machinations have been laid out for the world to see since the relevant vol-umes were published in the 1950s. Many of Jefferson's best-known letters from Paris had already been in the public domain for a century or more before that edition.

Jefferson and Adams arrived in Paris in the shadow of Benjamin Franklin, who was already there when they arrived. The three men formed a joint plenipotentiary commission "for negotiating treaties of commerce with for-eign nations" (Thomas Jefferson, *Autobiography*, in Peterson 1984, 54). Franklin was a world-class scientist, revolutionary, and showman. By the time that Jefferson and Adams arrived in Paris, he was already almost 80 years old, and not in very good health. His desire for an easy-going and cheerful life in the company of younger women did not please either Adams or Jefferson. Relations among the three plenipotentiaries were strained, and both the younger men were relieved when Adams was sent to London and Franklin returned home in summer 1785.

Franklin had nevertheless paved the way for his successor. As a member of the *Académie royale des sciences*, Franklin could introduce the eager amateur scientist Jefferson to Condorcet and his circle. Politically, Franklin and Jefferson were not close, but the distinctions of American politics eluded their French hosts. To the French, Franklin was a hero of the American Revolution, who had been denounced and insulted by the British after breaking with them. He had negotiated the American–French alliance. In his homely simplicity, he was assumed (wrongly) to be a Quaker. He was also assumed, also wrongly, to be the main author of the Constitution of Pennsylvania, which was widely studied in Paris.

A valuable primary witness here is John Adams. His personal copies of the two collections of U.S. Constitutions that were available in Paris at the time have survived. The first was a *Receuil des Loix Constitutives des Colonies Anglaises*. Translated by C.-A. Régnier, it was *dédié à M. le Docteur Franklin*, and purportedly published *à Philadelphie, et se rend à Paris* in 1778. (The Philadelphia imprint was almost certainly fake.) It contains the Declaration of Independence, and the constitutions of PA, NJ, DE, ND, VA, and SC. Adams was a great scribbler. In his books he maintains a continuous bad-tempered dialogue with the French Enlightenment. Much of it is transcribed in the seminal, but surprisingly neglected,[4] study by Haraszti (1952). At the start of the page containing the translation of the Constitution of Pennsylvania, Adams writes:

> The following Constitution of Pa, was well known by such as were in the secret, to have been principally prepared by Timothy Matlock, Jas. Gannon, Thomas Paine and Thomas Young, all ingenious Men, but none of them deeply read in the Science of Legislation. The Bill of Rights is taken almost

verbatim from that of Va.... The Form of Government, is the Worst that has been established in America, & will be found so in Experience. It has weakened that state, divided it, and by that Means embarrasses and obstructed the American Cause more than any other thing. (John Adams annotation in Adams Library, Boston Public Library, 233.7. My readings do not always coincide with Haraszti's 1952, 328)

This unicameral constitution of Pennsylvania is the target of Madison's attacks: overtly in *Federalist* 48, and more directly (although not by name) in *Federalist* 10 and 51. Jefferson fully agreed with Madison and Adams. Unfortunately, most of the French students of the American constitution, including Turgot, Condorcet, and the duc de la Rochefoucauld, were attracted by the constitution of unicameral Pennsylvania, backed by the supposed authority of the great Docteur Franklin.[5] La Rochefoucauld produced another translation of the U.S. state constitutions in 1783. It is more flowing than Régnier's and it includes explanatory notes. John Adams also had a copy of that edition, but did not annotate it, not even the section on MA (Van Doren [1938] 1991, 572; Adams Library, Boston Public Library, 40.2).

Adams and Jefferson—the two Americans to whom French constitution-writers turned for advice[6]—therefore had very mixed feelings about the American state constitutions. True, they were the authors of two of the seminal documents in the collection. Adams was the main author of the Constitution of Massachusetts (McCullough 2001, 220–25) and Jefferson of the Declaration of Independence. These facts, especially the second, were not widely known in Paris.

Jefferson's entrée to the world of science came especially via Condorcet. His entrée to the world of French liberal aristocratic politics came especially via Lafayette. Condorcet and Lafayette both tried to influence French discussion of human rights. Their circles intersected but were not the same. In the next two sections we study Thomas Jefferson's interactions with each. In summary: Jefferson and Condorcet were soul mates, Jefferson and Lafayette were not. Yet, through various contingencies, it was for Lafayette rather than for Condorcet that Jefferson "held the nail" that drove the Declaration into the French constitution.

JEFFERSON AND CONDORCET

Jefferson and Condorcet were men of very similar temperament, children of the Enlightenment who believed that science must banish human misery and superstition. Condorcet coined the term "sciences morales et politiques"; Jefferson may have been the first to refer to the latter in English as political science.[7] The mainspring of the moral and political sciences, according to Condorcet, was probability. The developing theory of probability had an extraordinary range of applications. It drove the new actuarial science and made stable insurance contracts possible. It powered Condorcet's jury theorem.[8] In a more oblique way it spurred him to produce the first axiomatic treatment of voting and majority rule. It informed his attitude to justice and human rights.

Condorcet was a professional scientist who used his position as Perpetual Secretary of the Academy of Sciences to control French and (as far as he could) European science policy. Jefferson was an enthusiastic amateur scientist. The final speech of the Heart to the Head acknowledges the Heart's respect for the Head's heroes: "Condorcet, Rittenhouse, Madison, La Cretelle, or any other of those worthy sons of science whom you so justly prize."[9] The respect was mutual. On Jefferson's side it was strained by the tragi-comedy of Citizen Genêt's mission to the United States in 1793 while Jefferson was secretary of state. Edmond Genêt was sent by the revolutionary French to stir up revolution in the United States, if necessary by appealing to the American people to rise up against cautious leaders such as President Washington or Secretary of State Jefferson. He was one of the most counterproductive envoys in history. Condorcet's last letter to Jefferson, endorsing Genêt's mission, may therefore account for Jefferson's temporary estrangement from Condorcet. But in one of the last documents he wrote in hiding before meeting his death in the Terror of 1794, Condorcet consigned his beloved daughter Eliza, should she escape to the United States, to the care of Jefferson, or of Franklin's grandson B. F. Bache. She did not reach the United States, but she and her mother Sophie de Grouchy survived the Terror. After Condorcet's death, if Thomas Jefferson's letter to William Green Munford of June 1799 quoted earlier is to be taken at face value, Jefferson was reconciled to Condorcet's values. In his wonderful post-1812 correspondence with John Adams (Cappon 1959), Jefferson never responded to Adams's fierce and frequent attacks on Condorcet and his fellow thinkers of the French Enlightenment. Adams thought that they were foolishly optimistic about human nature. Jefferson shared Condorcet's optimism.

Nevertheless, the intellectual relationship between Jefferson and Condorcet, both political theorists of the first rank, was not as fruitful as it might have been. Elsewhere (McLean and Urken 1992; McLean and Hewitt 1994), we have examined how much Jefferson or his lifelong collaborator Madison understood of Condorcet's revolutionary social science. Briefly:

- Jefferson understood Condorcet's probabilism. His letter to Madison, anthologised as "The earth belongs in usufruct to the living" (Peterson 1984, 959–64) derives both its formulae and its modes of reasoning from Condorcet, not, as the editors of the *Jefferson Papers* believed, (Boyd et al. 1950– , 15: 390 ff.) to Richard Gem.
- All Jefferson's holdings of Condorcet's works that survived until he sold the Monticello library to Congress in 1815 can be checked in the recently rediscovered catalog (Gilreath and Wilson 1989). We examined all that are known to survive (some were lost in a fire in 1851). Jefferson has some characteristically sharp annotations on his copy of Condorcet's posthumous *Esquisse d'un tableau sur le progrès de l'esprit humain* (*Outline of a*

Historical Picture of the Progress of the Human Mind, 1795). In particular he objects to Condorcet's claim that France was the first country to achieve religious freedom. No, says an angry Thomas Jefferson: Virginia was first. But he wrote nothing apart from his characteristic countersigning of the signatures[10] on his copies of Condorcet's work on voting theory. Adams, on the other hand, wrote an entire counter-manifesto in the margins of his copy of the *Esquisse* (Haraszti 1952, 241–56; Adams Library, Boston Public Library).

- Another intermediary between Condorcet and Madison was Philip (Filippo) Mazzei, a disreputable Italian-Virginian who wrote frequently to Madison and Jefferson (usually asking for money or to help settle suits against him; see *TJP; JMP;* Marchione 1975). Jefferson commissioned Mazzei to write a four-volume *Recherches Historiques . . . sur les Etats-Unis* in order to counter anti-American propaganda in Paris (much the same motive as for publishing his own *Notes on Virginia*). Mazzei (or Jefferson) inserted four chapters by Condorcet into this book, which Mazzei sent to Madison, unsuccessfully asking Madison to arrange a translation.

- Condorcet's four chapters were called *Lettres d'un bourgeois de New Haven à un citoyen de Virginie.* Condorcet was indeed a *bourgeois de New Haven*—he was one of ten distinguished Frenchmen made "Freeman of New Haven" at a town meeting in 1785. The *citoyen de Virginie* was Mazzei.

- These New Haven Letters argue for a unicameral national legislature, with representatives selected by a very complicated procedure.

- Madison refused Mazzei's request to get them translated, saying "I could not spare the time [and] . . . I did not approve the tendency of it . . . If your plan of a single Legislature etc. as in Pena. were adopted, I sincerely [*sic*] believe that it would prove the most deadly blow ever given to republicanism" (Madison to Mazzei, December 10, 1788, Hutchison et al. 1962– , 11: 388–89; see also Madison to Mazzei October 8, 1788, ibid., 11: 278–79.)

- John Adams had an even lower opinion of the New Haven Letters. In an 1815 letter to Jefferson, he wrote of Condorcet and the other *philosophes,* "These Phylosophers have shewn them selves as incapable of governing mankind, as the Bou[r]bons or the Guelphs. Condorcet has let the Cat out of the Bag." (John Adams to Thomas Jefferson, June 20, 1815, in Cappon 1959, 445. All of Adams's other references are equally derisive.) The New Haven Letters were the occasion of Adams's defence of bicameralism, *A Defence of the Constitutions of Government of the United States of America,* which he wrote in London in a great hurry in 1787 and immediately sent to Jefferson in Paris (McCullough 2001, 374–79; Adams Library, Boston Public Library, 131.12).

Jefferson in Paris took a very cheerful view of Shays' Rebellion in western Massachusetts in 1787. Whereas this rebellion against the independent government scared politicians in the United States sufficiently to give momentum to the Constitutional Convention, Jefferson insouciantly pointed out:

> We have had 13. states independent 11. years. There has been one rebellion. That comes to one rebellion in a century & a half for each state. What country before ever existed a century & half without rebellion? ...What signify a few lives lost in a century or two? The tree of liberty must be refreshed from time to time with the blood of patriots & tyrants. It is it's natural manure. (Thomas Jefferson to William Stephens Smith (Adams's son-in-law), November 13, 1787; Peterson 1984, 910–12)

Jefferson's language and his (dubious) statistical inference both come direct from Condorcet, who had written,

> In the eleven years that the thirteen American governments have existed, there has only been one uprising.... Imagine that the same thing occurred after the same interval in each of the other states. For an uprising to have taken place in all of them, we would have to wait 143 years. Under what other form of government are uprisings so rare? (Condorcet, *De l'influence de la Révolution d'Amérique sur l'Europe, Supplément*, 1787, translated by Sommerlad and McLean 1989, 289)

Jefferson admired Condorcet's mathematics much more than his politics. Condorcet's fatal error, in the eyes of all three of his American contemporaries Jefferson, Madison, and Adams, was to endorse unicameralism, and even the Pennsylvania constitution. Jefferson was not as doctrinal a bicameralist as either Madison or Adams, but he had made his feelings known in his *Notes on Virginia*. He had brought these *Notes*, originally drafted as replies to a set of queries from a French diplomat, with him to Paris, and he first published them there as part of the campaign to recruit French intellectuals to the American revolutionary ideology. Query XIII of the *Notes* contains Jefferson's striking denunciation of the "173 despots" who had replaced the solitary despot George III in the first Virginia constitution after independence. Although bicameral, "the senate is, by its constitution, too homogeneous with the house of delegates. Being chosen by the same electors, at the same time, and out of the same subjects, the choice falls of course on men of the same description.... An *elective despotism* was not the government we fought for" (Peterson 1984, 244–45).

Hence, although Condorcet and Jefferson had very similar ideas of human rights, it was not via Condorcet but via Lafayette that Jefferson chose to drive the nail home.

JEFFERSON AND LAFAYETTE

Lafayette admired Jefferson (not as much as he admired Washington, for whom his adulation is rather creepy). Jefferson did not admire Lafayette. However, he found him useful. Ample evidence for both points is scattered through the Jefferson Papers, but French constitution-writers do not seem to have noticed.

The 19-year-old Marquis de Lafayette, scion of one of the best-connected families of France, volunteered for Washington's Continental Army in 1776. Washington made him a major general. Jefferson met him first in 1781, when Lafayette commanded the force that delayed, but did not prevent, the British raid on Richmond and Monticello that forced Governor Jefferson to flee his state capital and his home, and cost him over 30 slaves freed by the British. Lafayette left the United States a hero (notably in his own eyes) and returned there for a victory tour in 1784. He was one of Jefferson's first French contacts on the latter's arrival. Jefferson presented him with a copy of the *Notes on Virginia* inscribed to one "whose services to the American Union in general & to that member of it particularly which is the subject of these Notes . . . entitle him to this offering" (quoted in Gottschalk 1950, 203).

Lafayette was no political theorist. He later constructed a myth of himself as the pioneer republican, but Gottschalk (1950, ch. 1) has shown that this was retrospective. Jefferson gave his view of Lafayette in letters to Madison:

> I find the M de la Fayette so useful an auxiliary [in Thomas Jefferson's trade negotiations] that acknowledgements for his cooperation are always due. (December 16, 1786; Boyd et al. 1950– , 10: 602)

> The *Marquis de La Fayette* is a most valuable *auxiliary to me*. His *zeal* is unbounded, & his *weight* with those in *power, great*. His *education* having been merely *military, commerce* was an unknown field to him. But his good sense enabling him to *comprehend* perfectly whatever is *explained to him, his agency* has been very *efficacious. He* has a great deal of *sound genius*, is well *remarked* by the Kin*g, & rising in* popularity. He *has* nothing against *him, but* the *suspicion* of *republican principles*. I think he will one day *be of* the *ministry*. His foible is, *a canine appetite for popularity and fame*; but he will get *above* this. (January 30, 1787; Peterson 1984, 885. Italicized passages sent in code)

Lafayette was thus the ideal tool for Jefferson's interests as they broadened from American trade to French politics. Jefferson was a remarkably undiplomatic diplomat. As the Assembly of Notables, the first step (as in turned out) on the road to revolution, prepared to assemble, Thomas Jefferson briefed Lafayette, who was of course to be a member:

> I wish you success in your meeting. I should form better hopes of it if it were divided into two houses instead of seven. Keeping the good model of your neighboring country [i.e., Britain] before your eyes you may get on step by step towards a good constitution. . . . The king, who means so well, should be encouraged to repeat these assemblies. You see how we republicans are apt to preach when we get on politics. (February 28, 1787; Boyd et al. 1950– , 11: 186)

If intercepted by government spies, this would hardly imperil Jefferson's position. But he became less and less cautious. We return to his tutoring of Lafayette in republicanism in the section on Jefferson and the French Revolution.

JEFFERSON AND THE U.S. CONSTITUTION

Jefferson was in Paris, not in Philadelphia, in 1787. Nevertheless, he had a substantial role in shaping the U.S. Constitution. As "Author...of the Statute of Virginia for religious freedom" (according to his self-written epitaph, Peterson 1984, 706–07), he played an important, albeit indirect, role in the entrenchment of the First Amendment to the Constitution. Together with his equally indirect role in the DDHC, it is his main contribution to both constitutional design and political theory. Both episodes illustrate the elusiveness of Jefferson that every commentator discovers.

In this as in most things he was close to James Madison. Madison and Jefferson had worked together in Virginia. Their proudest achievement was the Virginia Declaration of Religious Freedom. For the tortuous history of that document see Rakove 1990, 6–14. Jefferson's pride in it equalled Madison's. As noted, it led him to complain that Condorcet's *Esquisse* wrongly credited France, not Virginia, for pioneering religious freedom.

The Virginians were more radical on state and church than were the New Englanders. Adams's 1780 Constitution of Massachusetts still recognised the role of the town church as guardian of public order and social control. (By 1820 Adams had changed his mind, but his attempts to disestablish the church in the MA constitutional convention failed—John Adams to Thomas Jefferson, February 03, 1821, in Cappon 1959, 571–72.) By contrast, no one church was dominant in revolutionary Virginia. Madison had cleverly formed a coalition of dissenters to complete the disestablishment of the Episcopalian church there.

When Jefferson saw the Constitution as reported out of the convention at Philadelphia, he had two vociferous objections to it, which he repeated to several correspondents:

> I will now add what I do not like. First the omission of a bill of rights providing clearly & without the aid of sophisms for freedom of religion, freedom of the press, protection against standing armies, restriction against monopolies, the eternal & unremitting force of the habeas corpus laws, and trials by jury in all matters of fact.... Let me add that a bill of rights is what the people are entitled to against every government on earth, general or particular, & what no just government should refuse, or rest on inferences. The second feature I dislike, and greatly dislike, is the abandonment in every instance of the necessity of rotation in office, and most particularly in the case of the President. Experience concurs with reason in concluding that the first magistrate will always be re-elected if the Constitution permits it. He is then an officer for life. (Thomas Jefferson to Madison, December 20, 1787, in Peterson 1984, 916)

Jefferson's first objection—the absence of a Bill of Rights—was widely shared. It became clear to the Federalists—that is, to those in favor of ratifying the Philadelphia constitution—that they would not get the required nine states to ratify unless they promised to consider adding a bill of rights in the first Congress (Riker 1996, 203–28). Several reluctant ratifiers, including

NH, MA, and VA, attached clauses for the bill that they wished to see added. A committee chaired by Madison in the first House considered the proposed clauses. Madison's committee recommended 12 amendments, of which 10 were ratified and became the U.S. Bill of Rights. The religious section of the First Amendment was one of several on Jefferson's list that was ratified, and in substantially the words of the VA Declaration of Religious Freedom.

Jefferson's second objection, to the absence of term limits especially for the Presidency, set him at odds with Lafayette. Lafayette was the president of the French chapter of the Society of the Cincinnati. This was a veterans' organisation for Revolutionary War officers, whose president was George Washington. Jefferson and other republicans were deeply suspicious of the Society. They saw it as the nucleus of an American aristocracy, with Washington at its head, set to become the first monarch of the United States. They were even more alarmed when it was proposed that membership of the Society should be hereditary (Gottschalk 1950, 54–64). As it turned out, however, Washington settled the issue in his own way by retiring voluntarily, to general surprise, after his second term in the Presidency.

JEFFERSON AND THE FRENCH REVOLUTION

While thus trying to influence his own country's constitution, Jefferson was drawn more and more into reforming that of the country to which he was accredited. He gradually became less and less cautious. Though his intellectual soul mate was Condorcet, his chosen instrument was Lafayette, for whom he conducted, in Gottschalk's (1950, 374) happy phrase, an "informal seminar on political theory." In December 1788, with a second Assembly of Notables due to work out the arrangements for the forthcoming Estates-General, Thomas Jefferson wrote to a fellow Virginian, "All the world is occupied at present in framing, every one his own plan of a bill of rights" (Thomas Jefferson to James Currie, December 20, 1788, Boyd 1950– , 14: 366). In this section, we compare four such declarations:

1. Jefferson's own, sent to Lafayette and to the Protestant pastor and politician Rabaut de S. Etienne on June 3, 1789;
2. that of Condorcet (for the complex provenance of which see McLean and Hewitt 1994, 55–63);
3. the second of Lafayette's three efforts, composed in June 1789;
4. the DDHC as finally approved by the National Convention.

Table 1.1 attempts to set out the most important points from the four declarations with clauses on the same subject on the same row. The numbers of clauses are given when either the original is numbered or numbers can easily be assigned.

In June 1789, Jefferson is still in his cautious phase as to content, though not as to behavior. On the day of his letters to Rabaut and Lafayette, we have a witness statement from Gouverneur Morris. Morris, one of the main

Table 1.1 Four Declarations of Rights, 1789

Thomas Jefferson draft	*C draft*	*LaF 06/89 draft*	*Decn. as adopted*
The States General shall assemble uncalled . . . annually. The States General alone shall levy money on the nation.		(6) The principle of all sovereignty resides imprescriptibly in the nation.	(3) The source of all sovereignty lies essentially in the nation.
Laws shall be made by the States General only.		(8) The legislative power must be exercised essentially by deputies chosen in each district by free, regular, and frequent election.	
Habeas Corpus	Punishment after due process only.	(5) No man may be subjected to any law not previously approved either by him or his representatives and correctly applied.	[Covered in Art. 6–8]
Military subordinate to civil authority.	Military subject to civil legal procedures.		
Printers liable to prosecution only for publishing false facts.			
Abolish "pecuniary privileges and exemptions."		(1) Nature has made men free and equal; social distinctions may only be based on public utility.	(1) All men are born and remain free and equal in their rights. Social distinctions may only be based on public utility.
Honor old regime debts.			
	Death penalty for murder only. No torture. Fixed terms for judges. Defendant to have right of peremptory juror challenge. No standing army. All legal proceedings in public.		

Table 1.1   Continued

Thomas Jefferson draft	C draft	LaF 06/89 draft	Decn. as adopted
	Freedom to practise any profession. "The legislature may not prohibit any action which is not contrary to the rights of others or of society."	(2) Every man is born with inalienable rights, including the right to property, the right to honor and life, the complete ownership of his person and products and to resist oppression.	(2) The final end of every political institution is the preservation of the natural and imprescriptible rights of man. Those rights are liberty, property, security, and resistance to oppression.
		(3) The only limits to the exercise of each man's natural rights are those which secure to society the enjoyment of the same rights.	(4) The only limits to the exercise of each man's natural rights are those which secure to other members of society the enjoyment of the same rights. These limits may be fixed only by law.
	Everyone is free to follow whichever religion he sees fit.	(4) No man should be harassed for his religion, his opinions, nor for communicating his thoughts, providing they are not libelous.	(10) No one may be persecuted for his opinions or creed, provided that their expression does not disturb the public order provided for by the law.
		(7) The sole end of all government is the common good; the legislative, executive, and judicial powers must be distinct and defined; no body nor individual may have authority that does not emanate expressly from the nation.	(6) Statute law is the expression of the general will.
	No conscription. No right of fathers to punish wives, or children over 16. Freedom of press, association, and religion. Protection of public goods "such		

Table 1.1 Continued

Thomas Jefferson draft	C draft	LaF 06/89 draft	Decn. as adopted
	as scenery, rivers and so on." No taking of property without compensation. Women not to be disadvantaged by inheritance laws.	(13) As the progress of enlightenment, the introduction of abuses, and the rights of succeeding generations necessitate the revision of every human work, there must be provision for a constitutional convention.	
	…and many, many others.		

Sources

Column 1 Thomas Jefferson to Rabaut de S. Etienne, June 3, 1789. Boyd et al. 1950– , 15 : 165–68.
Column 2 Declaration of Rights "par le Marquis de Condorcet, traduite en Anglois par le Docteur Gem avec l'original à coté" (Thomas Jefferson's annotation). McLean and Hewitt 1994, 255–70. Our translation and paraphrase of the French text.
Column 3 Boyd et al. 1950–, 15: 230–33. My translation.
Column 4 Finer 1979, 269–71.

draftsmen of the U.S. Constitution at Philadelphia, had arrived in Paris to negotiate on behalf of the (unrelated) Robert Morris American tobacco monopoly with the French Farmers-General. He did not share the republican optimism of Jefferson, whom he was later to follow as American minister in Paris. Morris recorded in his diary:

> Go to Mr Jefferson's. Some political conversation. He seems to be out of Hope of anything being done to Purpose by the States General. This comes from having too sanguine Expectations of a downright republican Form of Government. (Diary for June 3, 89 in Davenport 1939, i: 104)

However sanguine his expectations, Jefferson's draft for Rabaut and Lafayette addresses mostly issues that, for the United States, were in the original document reported from Philadelphia, rather than the Bill of Rights. In June 1789 France does not yet have a constitution, still less a Bill of Rights.

Matters moved fast, however. Condorcet's Bill of Rights goes much further than any other document in this set. It is probably a little earlier than Jefferson's, but more far-reaching. Condorcet had failed to gain election to the Estates-General, but his political views were moving rapidly to the left. He was the only thinker of the Enlightenment to suggest that women should have equal rights with men, and it will be noted that he also includes very modern-sounding environmental rights in his list. If he did not get

Condorcet's list direct from him, Jefferson got it from his personal physician Richard Gem.

However, Condorcet was not only out of power, he was too radical for Jefferson's purpose. Jefferson did not believe that France, only just abandoning feudalism, was ready for a declaration of rights as thoroughgoing as he was pressing on the U.S. ratifying states. Comparing Lafayette's first draft with one of Gem's, the latter clearly influenced by Condorcet, Jefferson told Madison (Boyd et al. 1950– , 14: 438–39, January 12, 1789) that Lafayette's declaration was "adapted to the existing abuses." By the end of June or the beginning of July 1789, Lafayette produced a second draft of his bill of rights. It contains some phrases which certainly arise from his seminars with Jefferson, such as the first part of clause 1 "Nature has made men free and equal." An extremely Jeffersonian clause is Lafayette's # 13:

> As the progress of enlightenment, the introduction of abuses, and the rights of succeeding generations necessitate the revision of every human work, there must be provision for a constitutional convention.

Jefferson's most famous statement on the rights of succeeding generations is his letter to Madison, written just as he was leaving Paris on September 6, 1789. In it he proposes "on this ground which I suppose to be self evident, 'that the earth belongs in usufruct to the living,' that the dead have neither powers nor rights over it." As the probability that at least one of any pair of contractors has died reaches 50 percent between 18 and 19 years, Jefferson proposes that all contracts, including constitutional contracts, should be void after this time (Peterson 1984, 959–64). This idea (itself derived from Condorcet—see McLean and Urken 1992)[11] did not appeal to Madison, after his year's labours preparing for the Constitutional Convention, attending it, and campaigning for the constitution. Nor did it appeal to the French convention. This clause of Lafayette's was not adopted.

Others are addressed to French conditions in unJeffersonian terms (such as the second part of clause 1, "social distinctions may only be based on public utility"). Others again are very remote from Jefferson and seem to owe more to Rousseau, such as clause 6: "The principle of all sovereignty resides imprescriptibly in the nation."

Many hands worked on the DDHC. Lafayette opened the agenda by presenting a draft, based on his June draft but with minor changes following his further discussions with Jefferson. Several bureaus of the National Assembly produced drafts and the final text was a melange of drafts from different bureaus. Most of it was taken from the 6ᵉ Bureau, on which Lafayette did not sit. However, one member recorded:

> After comparing the various plans of a Declaration of Rights with that of M. de La Fayette, I observed that the latter is the text to which the others form merely a commentary. (Abbé Bounefoy, *Archives parlementaires*, August 19, 1789, quoted by Fauré 1990; my translation)

The most momentous difference is that the Declaration as adopted contains the Rousseauvian clause 6: "Statute law [Fr: *La Loi*] is the expression of the general will." This has been taken throughout French history until 1971 to mean that *La Loi*, expressing as it does the general will, is superior to any constitutional text, even the sacred Declaration of 1789. The National Assembly decided not to make the Declaration itself part of the Constitution. This may have been a blessing in disguise, in that all French constitutions until the Third Republic were short-lived. However, the 1789 Declaration was incorporated into the preamble of the Constitutions of both the 4th Republic (1946) and 5th Republic (1958 and current).

In 1946 and in 1958, there was no tradition of judicial review in France. None of the main political forces at either juncture—the Socialists, the Gaullists, the Catholic center, nor the Communists—was prepared to countenance nonelected judges interpreting the constitution or striking down legislation. Thus the force of the 1789 Declaration was symbolic. Indeed it coexisted in both constitutions with the very different and partly incompatible social and economic rights inserted by the left in 1946. The body that did have the power to review the constitution in the 5th Republic is deliberately not called a court (it is the *Conseil Constitutionnel*) and its membership was skewed toward the Gaullists. Nevertheless, this was the body that instituted judicial review in France, first striking down legislation in 1971, and doing so extensively when the Socialists attempted to nationalise various entities after 1981. In the latter set of rulings, the *Conseil* explicitly privileged the (Jeffersonian) "sacred right to property" from 1789 over the right (or even duty) to nationalise that appears in the 1946 preamble (for more details see Stone 1992; Stone Sweet 2002). Jefferson's tutorials for Lafayette continue to affect everyday life in France two centuries later.

That Jefferson was a man of deep contradictions on human rights is undeniable. It was obvious enough to his French hosts, who knew that a slaveholder was chiding them for their backwardness in adopting the rights of man—a slaveholder who thought that they were not ready for a declaration of rights as advanced as that of Virginia. The depths of Jefferson's thoughts are in his letters, where his beautifully expressed contradictions are laid out. The man who wrote his wonderful valedictory letter of June 24, 1826 to Roger C. Weightman, Mayor of Washington DC, also wrote the wonderful phrases but nihilistic politics of the April 22, 1820 letter to John Holmes (Peterson 1984, 1433–35 [Holmes]; 1516–17 [Weightman]).

In the Holmes letter he described the proposal to ban slaveholding in the State of Missouri as "a fire bell in the night. . . . I considered it at once as the knell of the Union . . . we have the wolf [of slavery] by the ears, and we can neither hold him, nor safely let him go. Justice is in one scale, and self-preservation in the other." Jefferson never saw his way out of that dilemma. Whether or not he had children by his slave Sally Hemings (and the case is not proven, despite Foster's [1998] misleading title[12]), he was certainly paralysed by the knowledge that slavery contradicted the principles of human rights, but he could not contemplate a State of Virginia without it.

Declining Weightman's invitation to attend the July 4th celebrations in Washington DC on grounds of health, the dying Jefferson wrote:

> May it [the Declaration of Independence] be to the world, what I believe it will be, (to some parts sooner, to others later, but finally to all,) the signal of arousing men to burst the chains under which monkish ignorance and superstition had persuaded them to bind themselves, and to assume the blessings and security of self-government. That form which we have substituted, restores the free right to the unbounded exercise of reason and freedom of opinion. All eyes are opened, or opening, to the rights of man. The general spread of the light of science has already laid open to every view the palpable truth, that the mass of mankind has not been born with saddles on their backs, nor a favored few booted and spurred, ready to ride them legitimately, by the grace of God.

In one of his best acts of literary detection, Douglass Adair (1974, pp. 192–202) showed that the image of "saddles on their backs" comes from the dying speech of Col. Richard Rumbold, a former Cromwellian sentenced to death for rebellion against the Catholic King James II in 1685. This is Jefferson the opposition Whig, like so many of the American revolutionaries seeing the revolt against the British Crown as the country against the Court. But the rest of the imagery is distilled Enlightenment thought. "Monkish ignorance and superstition" is pure Voltaire, probably mediated through Condorcet's *Esquisse*. "All eyes are opened, or opening, to the rights of man" recalls 1789 even more than 1776. Jefferson in Paris was not just a movie, but a seminal event in the history of human rights.

NOTES

1. Benjamin Franklin, Jefferson's predecessor as minister in Paris, dallied with Mme Helvétius at her salon in Auteuil. When Jefferson and John Adams arrived to join Franklin in Paris, they both disapproved of his behavior. In 1786, the young Sophie de Condorcet held a salon in her husband's apartment at the Hotel des Monnaies, quai de Conti (opposite the Louvre: Guillois 1897, 68–76). After the Terror of 1793–94 and the death of her husband, Sophie moved into Mme Helvétius's old house at Auteuil and reopened her salon (Guillois 1897, 94, 177). I read this as a defiant statement of her radicalism and feminism.
2. Conor Cruise O'Brien, in his recent controversial *The Long Affair: Jefferson and the French Revolution* (1996), denies that Jefferson was ever close to Condorcet or to any other French Enlightenment figure. He also claims that Jefferson never learnt French. A quick scan of the Princeton edition of the *Jefferson Papers* easily disproves these claims.
3. "The Girl she [Polly Jefferson] has with her, wants more care than the child, and is wholly incapable of looking properly after her, without some superiour to direct her" (Abigail Adams to Thomas Jefferson, July 6, 1787, in Cappon 1959, 183).
4. Even by David McCullough, whose acclaimed biography (McCullough 2001) has single-handedly put Adams back in the pantheon where he belongs.

McCullough cites Haraszti, but barely uses him. The custodians of Adams's books in Boston Public Library told me in December 2001 that demand to read them had scarcely risen since McCullough (2001) had been published.

5. Franklin was rarely present at the PA constitutional convention of 1776, which he nominally chaired. But he did approve of unicameralism, see the letter quoted by Van Doren [1938] 1991, 554.

6. Tom Paine, a principal author of the PA constitution according to John Adams, was in Paris in 1787 and again in 1789–90. But he spoke no French. On the first visit, he was mostly promoting his iron bridge. On the second, although he met Lafayette, there is no strong evidence that he influenced the DDHC.

7. Another claimant is Alexander Hamilton.

8. Condorcet (1785). The jury theorem states that the probability that a decision is correct is a positive monotonic function of two things: the average enlightenment of the jurors, and the size of the majority. After two centuries of neglect, it is once again at the centre of scholarly attention. See Austen-Smith and Banks 1996; Miller 1997; List and Goodin 2001.

9. By Madison, Jefferson probably meant not the politician but his cousin and namesake Rev. James Madison, president of William & Mary College. Jefferson called the Philadelphia scientist David Rittenhouse "second to no astronomer living; . . . in genius he must be the first, because he is self-taught."

10. Every 16 or 32 pages, a book had a consecutive letter in the bottom margin to show the binder in which order to bind the pages. These marginal letters are known as "signatures." Jefferson marked his ownership of books by writing a "T" before signature J, and a "J" after signature T.

11. Also, it has to be said, an idea that must have attracted Jefferson personally, who was in chronic debt from the moment he took on the liabilities of his father-in-law's estate in 1778 until the day he died with liabilities hugely in excess of his assets. See Sloan 1995.

12. Foster et al. have proven beyond reasonable doubt that *a* Jefferson fathered Sally Hemings's last child. Five Jeffersons with the marker Y-chromosome haplotype were alive when Eston Hemings was born. The case that Thomas Jefferson was the father is persuasive but not conclusive.

REFERENCES

Adair, D. (1974). *Fame and the Founding Fathers*. Ed. T. Colburn. New York: Published for the Institute of Early American History and Culture at Williamsburg, VA., Norton.

Austen-Smith, D. and Banks, J.S. (1996) "Information aggregation, rationality, and the Condorcet jury theorem." *American Political Science Review* 90 (1): 34–45.

Boutmy, E. (1902). "Review of Jellinek 1902." *Annales des Sciences Politiques* 415–53.

Boyd, J.P. et al., eds. (1950–), *The Papers of Thomas Jefferson*. Princeton: Princeton University Press. Main series 28 vols to date.

Cappon, L.J. ed. (1959) *The Adams – Jefferson Letters: The Complete Correspondence Between Thomas Jefferson and Abigail and John Adams*. Chapel Hill: University of North Carolina Press.

Colliard, C.-A., ed. (1990) *La Déclaration des droits de l'Homme et du Citoyen de 1789: ses origines, sa pérennité*. Paris: La Documentation française.

de Condorcet, M.J.A.N. (1785) *Essai sur l'application de l'analyse à la probabilité des décisions rendues à la pluralité des voix.* Paris: Imprimerie Royale.

Davenport, B.C., ed. (1939) *Gouverneur Morris : A Diary of the French Revolution.* 2 vols. Boston: Houghton Mifflin.

Fauré, C. (1990) "La Déclaration des Droits de 1789 : le sacré et l'individuel dans le succès de l'acte" in C.-A. Colliard ed. La Déclaration des droits de L'Homme et du Citoyen de 1789: ses Origines, sa Pérennité. Paris: Le Documentation française. 71–79.

Finer, S.E. (1979) *Five Constitutions.* Harmondsworth : Penguin.

Foster, E. et al. (1998) "Jefferson Fathered Slave's Last Child." *Nature,* November 5.

Gilreath, J. and Wilson, D.L. (1989) *Thomas Jefferson's Library: A Catalog with the Entries in His Own Order.* Washington DC: Library of Congress.

Gottschalk, L. (1950) *Lafayette Between the American and the French Revolution.* Chicago: University of Chicago Press.

Guillois, A. (1897) *La marquise de Condorcet: sa famille, son salon, ses amis, 1764–1822.* Paris: Paul Ollendorf.

Haraszti, Z. (1952) *John Adams & the Prophets of Progress.* Cambridge, MA : Harvard University Press.

Hutchinson W.T. et al., eds. (1962–) *The Papers of James Madison.* Main series 17 vols to date. Vols 1–10 publ. by Chicago University Press, thereafter by University of Virginia Press.

Jellinek, G. (1902) "La Déclaration des droits de l'homme." *Revue du Droit Politique* 18: 385–408.

List, C. and Goodin, R.E. (2001) "Epistemic Democracy: Generalizing the Condorcet jury theorem." *Journal of Political Philosophy* 9 (3): 277–306.

Malone, Dumas (1951) *Jefferson and the Rights of Man.* Life of Thomas Jefferson, vol. 2. Boston: Little Brown.

Marchione, M. (1975) *Philip Mazzei: Jefferson's "Zealous Whig."* New York: American Institute of Italian Studies.

McCullough, D. (2001) *John Adams.* New York: Simon & Schuster.

McLean, Iain and Urken, A.B. (1992) "Did Jefferson or Madison Understand Condorcet's Theory of Social Choice?" *Public Choice* 73: 445–57.

McLean, Iain and Hewitt, Fiona (1994) *Condorcet: Foundations of Social Choice and Political Theory.* Cheltenham: Edward Elgar.

Miller, G.J. (1997) "The Impact of Economics on Contemporary Political Science." *Journal of Economic Literature* 35 (3): 1173–204.

Peterson, M.D., ed. (1984) *Thomas Jefferson: Writings.* New York: Library of America.

Rakove, Jack N. (1990) *James Madison and the Creation of the American Republic.* New York: Longman.

Riker, W.H. (1996) *The Strategy of Rhetoric.* New Haven: Yale University Press.

Shepsle, K.A. and Bonchek, M.S. (1997) *Analyzing Politics.* New York: W. W. Norton.

Sloan, Herbert (1995) *Principle and Interest: Thomas Jefferson and the Problem of Debt.* New York, Oxford : Oxford University Press.

Sommerlad, Fiona and McLean, Iain (1989) *The Political Theory of Condorcet.* Oxford University Social Studies Faculty Centre Working Paper 1/89.

Stone, Alec (1992) *The Birth of Judicial Politics in France.* Oxford: Oxford University Press.

Stone Sweet, Alec (2002) "Constitutional courts and parliamentary democracy." *West European Politics* 25 (1): 77–100.

Van Doren, C. (1938, 1991) *Benjamin Franklin.* New York: Penguin Books.

CHAPTER 2

LIBERAL IMPERIALISM

Alan Ryan

In this chapter, I argue that liberalism *is* intrinsically imperialist; and that we should understand the attractions of liberal imperialism, and not flinch. However, I argue against succumbing to that attraction. This is for low reasons of practicality rather than on high moral principle, though "practicality" has considerable moral force when people's lives are at stake. The paper sustains an argument for a nonmilitary but morally uninhibited global liberalism, *if it were pursued intelligently and consistently,* and for that I offer no apology. I regret the topicality of the subject—I have been bothered by it in the terms set by this paper since the Suez and Hungarian invasions of the Fall of 1956.

I also regret that I have not been able more elegantly to integrate my few thoughts about Jefferson into the broader argument, and have therefore inserted them as a sort of intermezzo. Jefferson is for my purposes altogether too good an illustration of the temptations to which liberals can succumb, as well as an illustration of the very unusual conditions that have to obtain if the liberal imperialist is to be successful. However, he has been so variously assaulted for his views on slavery, his treatment of native Americans, and a great deal else that a short essay on the pros and cons of liberal imperialism—or idealist interventionism—cannot hope to do justice to Jefferson in passing. My last preambulatory point is this: philosophers argue extreme cases, which is good practice in the company of other philosophers but rhetorically unhelpful in other circumstances. The extreme cases in this essay serve a heuristic purpose: to make one think about what would have to be true for the case the essay makes to be compelling as a guide to practice. Anyone who dislikes the argument can think of it as a form of *reductio ad absurdum.*

I

Liberal imperialism, or liberal interventionism, is the doctrine that a state with the capacity to force liberal political institutions and social aspirations upon non-liberal states and societies is justified in so doing. Let us begin by noticing that contrary to the orthodoxy in political theory, there is nothing odd about the suggestion that we can force an individual or a society to be free. It is impossible to force either to "do freely" what we force them to do; doing it freely and doing it under compulsion rule each other out. That, however, does not settle the matter. Imagine a small colony, as it might be Malta under British rule; imagine the imperial power having tired of its responsibilities toward the inhabitants of this small colony, but the inhabitants not having tired of being a colony. If Britain were to walk away from her imperial role, she would have forced freedom on Malta; in fact, something not unlike this happened to Malta and to some of Britain's Caribbean colonies who found more self-government thrust upon them than they at that time desired. Were it not for the certainty that a savage civil war would have followed, the British government might have forced freedom on Ulster. You may be tempted to object that this is a stretch in the notion of forcing anyone to do anything: Britain did not "use force" in the ordinary sense, but rather refused to go on operating as the ruling authority. What one might describe as a "refusal to govern" may seem to have too few of the features associated with the employment of force in international affairs, to provide much illumination.

It is not intended to provide a lot of illumination, only to show that forcing communities and individuals to be free is not an incoherent notion. The refusal to govern forced its former beneficiaries to do without the previous assistance, and so forced freedom upon them. Take an individual case. A man may become habituated to life in prison and be unwilling to leave. The time comes when his sentence is completed and the prison wishes to be rid of him. He refuses to go; a struggle ensues, and he is pushed out the door. Under those conditions he has had his liberty forced upon him. Cases may be multiplied. A married man is not free to marry; as the unwilling victim of a divorce thrust upon him by his wife he becomes free to marry. He, too, has been forced to be free. In general, once one has identified *what* condition is it that constitutes being free in the appropriate respect, the question whether someone else can force that condition upon us is something to be answered empirically. You may say that the prisoner expelled from jail and forced into freedom need only break a window to have himself rearrested and reincarcerated, and that is true enough. We can force freedom upon him but we shall have a hard time keeping him in the condition into which we have forced him, a consideration that applies with equal force at the societal level. The victim of an unwilling divorce on the other hand may remain unhappily free to contract another marriage but without any inclination to do it, and find that his new-found freedom is just about inescapable.

Such examples push out of the way any suggestion that there is a *conceptual* barrier to liberal imperialism. There may be any number of objections of practicality and morality, but it cannot be said that the very suggestion of forcing societies and their members "into freedom" is sheerly incoherent. This is not an argument that Americans ought to resist. During the War of Independence, there was no settled majority in favor of breaking ties to Britain; it is more than likely that until late in the war most colonists were dubious about a total breach, unsure whether they could sustain a government of their own if they were to succeed, and none too certain that their immediate neighbors would be less inclined to tyranny than the distant George III and Lord North.

The methods employed by the patriots were not those that twenty-first-century American governments approve when employed by liberation movements in foreign countries—the doubters were driven from their homes and farms and threatened with death or injury if they returned. To these disadvantages of refusing to join in the revolution, there was added the positive incentive that adherence to the revolution was a quick way to acquire the farm and property of loyalist neighbors. By comparison, the African National Congress has all along behaved astonishingly well. No American will say that those who were forced *out* of dependence on the colonial rule of the British and subsequently *in* to the republican arrangements concocted by Madison and his friends were less free—politically speaking—at the end of the day than at the beginning. They might be poorer, injured in the war, unsure what the future would hold; but they were free. The change was less dramatic than advertised by those who instigated it, but there was and remains a well-understood sense in which republican institutions are paradigmatically the institutions of free government, and monarchical institutions of the British sort are not. Free institutions had been brought into the world, not just by dissenters from the old regime severing their ties to it but by forcing a sufficient number of grudging American colonists to join in, too. Out of mutual coercion comes freedom.

The cases that we nowadays have to think about, and the ones that raise problems in their most acute form are those which rest on the thought that liberal western nations ought to at least protect and where possible to spread liberal values, and to do it by brute force when necessary. I want to take two actual cases and one nineteenth-century text, and then draw some inferences from them. As always, my stalking horse is John Stuart Mill, a thinker who was not only the author of the essay *On Liberty*, but also the senior London-based civil servant in charge of the government of British India. Before he became Examiner of India Correspondence not long before he retired in 1858, Mill occupied the wonderfully ambiguous role of attending to the political arrangements of the "Princely States," the Indian entities that were governed by their own Maharajahs under the tutelage of a political officer of the East India Company (EIC). Since Mill was always admirably frank in acknowledging that the EIC exercised a "despotic" government over the sub-continent, he could have made some interesting observations on the nature of the indirect despotism that he himself had operated.[1] He did not.

Indeed, one of his more striking, though not wholly admirable, rhetorical gestures was to dismiss colonial issues from the opening pages of *On Liberty* with the observation that despotism is a legitimate method of governing barbarians, so long as their improvement is the aim of government and the result of its activities in fact. Queen Elizabeth—the First—along with Charlemagne and Akhbar, are among the beneficiaries of this somewhat broad-brush theory of government. We ought not to be too hard on Mill, because he in fact gave a very elegant justification of the despotic rule of the EIC when he appeared before the House of Lords Committee that was considering the last renewal of the Company's charter in 1852. Their Lordships were less shocked by Mill's assurances that the Company provided good government in spite of lacking the usual apparatus of checks and balances that secured good government in Britain, than by his insistence that as soon as the Company had taught Indians the arts of self-government, and had provided a proper infrastructure of law on the one hand and docks and railroads on the other, it would be time for the Company to pack up and leave. The self-eliminating imperial project was a cleverer political concept than their Lordships were accustomed to dealing with. As Mill feared, once the EIC was abolished in 1858, the British decided that the Raj was not a training ground for the arts of liberal self-government but the greatest jewel in the imperial crown and not to be given up lightly.

It is important, though Mill was not absolutely clear about it in front of the committee that Mill did not suggest that the British government or the EIC should patrol the world looking for countries that might be dragged willy-nilly into liberalism and representative government. In this he was much more cautious than Bertrand Russell, who several times suggested that instead of fighting each other, European nations should gang up and divide the world between themselves in order to civilize it. Mill thought that no nation has a right to civilize another against its will; what he says in *Liberty* about the U.S. view that a "civilisade" should be launched against the polygamist Mormons of Utah embodied that thought.[2] Three things are worth noticing. First, Mill thought that *whatever the originzs* of the EIC's control of the Indian sub-continent, there was a liberal account to be given of the way the EIC should now behave, given that it was in a position to act—and, indeed, given that once it was the only plausible power in the sub-continent a failure to act would plunge its Indian subjects into chaos, disorder, and exploitation that would be far worse for them than the rule of the EIC. The EIC may have had no right to be in India in the first place, but once there, it had to act. What is interesting about Mill's House of Lords evidence is that Mill held that *given* the EIC's situation, its proper course of action should not be dictated by the self-interest of Britain, the imperial power, but by the requirements for the political development of India, construed in a liberal fashion.

Second, Mill's view was that *even if* we had a right to go around civilizing people against their will, there would be a very strong case for not doing it. I shall elaborate on that in due course, but the obvious basis for Mill's assertion that we have *no* right to civilize anyone against their will is the same as the basis for the argument that even if we had such a right, we ought not to exercise it. If one person can take civilizing measures at his own discretion, so may any other person, and there will be chaos. This is a consequence of Mill's consequentialist and utilitarian view of rights, and fits in with what he says elsewhere about why each of us has the right not to be murdered even if the world would on balance be better off for our death. If one person can make that decision, then so may anyone, and the safety and predictability that is the object of any system of rights to promote will have gone down the drain.

Third, Mill is delicacy itself compared with Marx. Inheriting Hegel's view that history is a "slaughterbench" or in the alternative, a court of justice where the only available sentence is death, Marx thought that the EIC was certainly wicked, but no more so than any other capitalist enterprise. It went without saying that men were motivated by greed and the desire to accumulate wealth, and that the EIC was organized so as to bring home to Britain whatever could be extracted from India. Just as Hegel held that we were doing the work of God—or, more properly, *Geist*—when our wicked behavior propelled history toward the rational freedom of the modern world, so Marx held that the EIC was justified because it dragged its Asian victims into modernity. By the same token, he thought Bismarck justified in taking Schleswig-Holstein by force, and dealing reactionary old-Norse nationalism its deathblow. He was not inclined to flinch at the brutality with which Asian stagnation was overcome. The "Left" is not immune from the temptations to which Mill might be thought to have succumbed.

A second historical example is provided by World War II. There is no denying that the Axis powers were bombed into liberal democracy. There are ways of softening this thought. We may say that the German people were enslaved by Hitler and that bombing Germany into acceptance of the allies' view of the world was less a matter of bombing the non-liberal German people into a relatively liberal frame of mind than a matter of bombing Hitler and his friends out of their capacity to enslave the rest of their countrymen. Nobody thinks there is anything wrong with using force to free someone *other* than the target of that force. If you are held captive by a hostage taker and I contrive to hit him over the head and liberate you, I have employed force to free you, and I have freed you by force, but nobody would say that I have forced you to be free. Only if you had become beguiled by your captor so that when I immobilized him, you tried to remain where you were, would there be a question of my having to force you to be free. Only if you try to stay in the bank cellar where he had imprisoned you, and I had to drag you out by main force, would it be plausible to say that I was forcing you to be free.[3]

In fact, neither Germany nor Japan in 1945 was unequivocally—or in Japan's case even halfheartedly—a free society fallen into the hands of oppressors. Both were substantially committed to following their authoritarian leaders into the disastrous courses to which that obedience led. Certainly, eliminating the bad leaders was a crucial step toward liberation; that is, it was a large part of the movement toward freedom to disable those who could have prevented the citizenry from creating a free regime. It might well have been a crucial negative step in another sense, which is that by disabling rulers whose legitimacy rested on their ability to deploy force successfully, and on little else, the allied forces did more than literally disable the erstwhile rulers; they delegitimated militaristic regimes.

Since I am offering these thoughts in a somewhat "arm's-length" fashion, I ought to stress that I am unequivocally committed to the implications of that last sentence. In order to survive, a rights-respecting regime must show itself to be as capable of maintaining order and enforcing the rule of law as any of its illiberal competitors. This is not difficult as a general rule; in times of peace, liberal regimes have lower crime rates and better compliance with commercial and tax laws than their rivals. What is less obvious is that liberal regimes are in general as militarily effective as their illiberal competitors. Soldier for soldier, the German armies in World War I and World War II were perhaps 50 percent more effective than their American and British opponents, not because of individual courage so far as I know, but because of the quality of the military staff. What the Axis powers could not do, and their enemies could, was to mobilize the entire nation to throw everything into the war; nor could they sustain cooperation with allies who were dubious about the relationship or hostile to it. (Nor, for that matter, could the Soviet Union sustain effective alliances.) So, I do not want to suggest that it was a small matter that the Allies could put together the resources to smash the German war machine. If there were a choice between liberalism and effective government, life would be even more difficult than it already is.

It is central to the question that I raise—how far can nations with capacity to impose liberal institutions and allegiances go in doing so?—that the positive impact was more than the elimination of dictatorship: it was the compulsory installation of liberal democracy. Importantly, what was installed was at any rate modestly well-adapted to the local cultures—Japan became a parliamentary constitutional monarchy rather than a rip-roaring republic, for instance. Indeed, one of the most obvious reasons for being extremely wary of spreading liberalism at the point of a bayonet is just that it is extraordinarily difficult to insert new ideals into cultures that are very foreign to one's own. At least one element in the postwar success of the allies must have been the fact that six years and four years of total war had created something closer than usual to *tabula rasa* for the nation-building efforts of the Allies. Still, the locals were not asked what sort of arrangements they would like now that the bombing had stopped; they were presented with a new constitution and told to endorse it. More than that, they were occupied by the troops of the victorious power(s) until it was apparent that the liberal virus had sufficiently infected its host. The subsequent 57 years have thus far been a good advertisement for the allies' actions in the immediate postwar years. Italy, Germany, and Japan are not models of Jeffersonian virtue, nor can one imagine their populations (or anyone else's) staying up late to read Mill's essay *On Liberty*. Nonetheless, they are solidly liberal democratic societies, and they got that way by being bombed into that condition. That is the thought on which I want to insist. Many people dislike this sort of argument so much that they deny that any such events as these occurred. They would do better to agree that they occurred, but to argue that they are such special cases that they cannot bear the construction I put on them. So, I shall now show why they might be special cases, then why they may not be very special.

So, what reasons might lead us to think my two cases are so special that no general principles can be drawn from them? First, India. The danger, we agree, is to believe that any nation that thinks itself civilized may go around liberating whatever other nations it can bend to its purpose. Kant thought that no state had the right to civilize another state by force, and Mill sounded the same note in condemning the U.S. treatment of the Mormons. The concept of a "civilisade" was one he repudiated. How might we distinguish the Indian case and the Mormon case? Mill's argument was that the EIC had simply found itself in control of India. Nobody had decided to add the political responsibilities of government to the commercial undertakings of the EIC; but having been led step by step into a situation where the government had fallen into its hands, the EIC had only two choices—to govern well or to allow anarchy to prevail. Nobody had found themselves in control of the territory occupied by the Mormons, and what was being proposed was an invasion on the basis of pure moral principle.

Second, Germany and Japan. Leaving aside the technicality that Britain declared war on Germany rather than *vice versa*, World War II was a defensive war. Just as nobody could accuse me of forcing good behavior on the burglar whom I disarm, nobody can accuse the Allies of forcing liberalism on the aggressor nations they disarmed. As between teaching the burglar the error of his ways and leaving him to go on and burgle other people, one should surely do the former. It would be absurd and counterproductive to arrest everyone within sight who struck us as vulnerable to temptation, and lecture her or him on the evils of burglary. Having begun the war in self-defence we may legitimately end it by imposing liberal institutions, but its origins in self-defence are what legitimate our doing anything whatever. In short, if accident puts us in the way of *having* to act, or self-defence does the same, we may act; otherwise, we have no right to act. The consequentialist liberal is half impressed; he will want an argument that underpins talk of when we have a right to intervene, and will think that such an argument must be consequentialist and that if it is sound it supplies all we need without a diversion through a discussion of the rights of states. Consequentialists are inclined to think that appeals to "rights" are abbreviations of more complicated discussions of consequences, a necessary shorthand but no more; and it is from that standpoint that this essay is written.

Consequentialism has no room for the idea that situations are special. Once we have accepted that we are not going to patrol the globe looking for folk to turn into good liberals, we are little further forward with the question of how we are to conduct our interactions with the rest of the world. The consequentialist takes seriously the state of the world as it now is. Two centuries ago, one could have decided that the best the liberal could do was to mind his own business, and leave everyone else to mind theirs, but in an increasingly intermeshed world that is implausible. "Mind your own business" must now express an old-fashioned realism. Realism in this context is the injunction to ask only what we have to gain from any given interaction in the foreseeable future, ignoring everything that goes on within the borders of another state as none of our business. A government animated by realism will take its own citizens' moral commitments—liberal, conservative, pacifist, or militarist—into account to the degree that they influence its ability to pursue the national self-interest; but it will not take them seriously *as* moral commitments.

The liberal who is not-quite-an-imperialist cannot be a realist in this sense. The liberal must say something like the following. Realism is a perfectly possible policy; a state can operate in international affairs by asking only "what's in it for us?" Clean-hands abstentionism is not a possible policy for liberals. We may try to avoid violating the rights of other nations by avoiding interaction, but we shall fail; if we do not molest others, they will molest us, either by accident or design, or by way of third-party interactions, and we shall then have to form a policy towards them. In any case, abstentionism is not obviously clean handed. Should we let Ethiopia starve when we could help it with grain? Is this not too like ignoring the starving baby left on our doorstep? If grain, why not medical supplies? If medical supplies, why not a few peacekeepers so that the supplies can get to those who are injured in the latest civil war? If.... Either we pursue a policy of clear-eyed selfishness, or we open up the argument to the question of how much good we should aim to do more generally. To this, the only answer is that we should try to do as much good as possible, remembering all the while that we may need a great deal of self-restraint on those occasions when doing nothing is the best we can do. But if a liberal wants to do as much good as possible, then the good s/he wants to do must be colored by liberal ideas regarding what that good is.

II

This is the moment to contemplate Jefferson. On the face of it, Jefferson stands at the opposite end of the intellectual and political spectrum from anyone who is moved by the argument of this paper. His hostility to overseas entanglements was often and vigorously expressed; his desire that the United States should have no standing army and no navy was notorious; however it was that liberty might spread through the world, it was on its face, not by means of American aggression against the illiberal and the reactionary. Yet, Jefferson not only imagined that the United States might within 50 years be a nation of 50 million inhabitants, he did his best to provide it with the necessary space by the Louisiana Purchase. Nor was he prepared to acquire territory only by purchase. He threatened the native American peoples who stood in the way of endless expansion with "extermination" and envisaged seizing Canada in order to chase them into their last refuge and exterminating them there. From the point of view of the original inhabitants of the continental United States, he committed the United States to a war of conquest. The purpose for which he did it was to spread freedom and enlightenment. One critic who dislikes Jefferson's conception of enlightenment has compared him to Pol Pot.[4]

Curiously enough, at the same time that the nascent United States pushed westwards into the interior of the continent, the Czarist government of Russia began its nineteenth-century expansion eastward, to the Caucusus and beyond into Siberia. In terms of the prudential considerations that this essay is intended to keep in the forefront of our minds, both the Americans and the Russians were favored by one crucial fact—the territories into which they expanded most readily were sparsely populated by nomadic and semi-nomadic peoples at a much lower level of technological and political development than themselves. That fact, however, renders their expansion uninteresting for our present purposes. From the perspective of the liberal consequentialist, it goes without saying that it was better to bring the American continent under the control of a generally liberal, democratic, and rights-respecting regime than it would have been to have subjected it to an inefficient bureaucratic despotism. However, that leaves out of consideration the prior question whether the accumulation of territory is a legitimate goal, let alone the situation of the dispossessed aboriginal inhabitants. The peculiarity of Jefferson's behavior from the perspective of this essay is that much of his early reflection on politics suggests a commitment to what I previously called "clean hands abstentionism," as though it was enough for American to set about preserving freedom and self-government within the new United States and to leave the rest of the world alone. Reinforced by his prejudice that self-sufficient farmers were the salt of the earth and all other occupations corrupting to a greater or lesser degree, this view dismissed most trade as an "entanglement," and was hardly a basis for an aggressive liberalism that might take political salvation to other peoples sword in hand. Conversely, when Jefferson imagined a vastly expanded America, he dismissed many of the inhabitants of other countries as simply inapt for freedom. Indians who show no sign of turning into quiet farmers are not to be persuaded but "exterminated." This is infinitely far from Mill's idea of tutelary despotism, and by the same token infinitely far from the temptation that this essay is trying to make real, the temptation to use the military and other resources of a highly developed country to—and the issue we have not yet confronted is that of how far the process is to go—at least eliminate brutal and inhumane regimes where possible and replace them with something closer to liberal democracy, and perhaps engage in wholesale cultural reconstruction to provide the newly built regimes with adequate support.

The one element of Jefferson's lifelong views that we may want to recall at this point relates less to the liberalism of the liberal imperialist project than to its imperialism. Jefferson, like his educated contemporaries, thought about eighteenth-century America in the light of English history and a tradition of republican theorizing that went back to Cicero. In this tradition, there are always two things to remember about empire; the first is that it rarely does for those who are subjected to imperial rule quite what the imperialists have in mind, and the second that imperialism corrupts republican politics. Taking the first for granted—that is, taking it as read that the history of the past several centuries gives us few grounds for confidence that

even well-intentioned imperial powers in fact know how to make things
happen as they hope—the second is worth focusing on for a moment.
Eighteenth-century Englishmen fearful of the effects of Indian wealth on
English politics were no doubt partly anxious only about the rising cost of a
seat in parliament, but behind that anxiety was an unease that went all the
way back to the last century of the Roman republic. Rich men accustomed
to lord it over the semi-slave inhabitants of whatever the latest conquered
territory might be, were not to be trusted in domestic politics when they
came home. In the twentieth century, one might have added to these anxieties the suspicion that the brutality of European rule in Africa had come
home to roost in the savagery and racism of European politics between 1914
and 1945. The prudent liberal imperialist will be almost as fearful as
Jefferson of the dangers of turning democracy into plutocracy by imperial
accident.

III

Leaving Jefferson aside, then, the remainder of the essay first of all defends
the attractions of what is often taken for wickedness before conceding much
of the case of the critics. We may start with a "maximalist" argument: the
duty to maximize the number of decent liberals in the world; and then move
to a more minimalist argument from the universal duty to defend anyone's
fundamental rights when they are threatened. The liberal's answer to the
question of what is the most good we can do in international affairs is that
we should do our best to promote liberal values and liberal institutions. One
(and only one) interpretation of this is that we should try to maximize the
number of committed liberals in the world, since they will create and sustain
the liberal institutions that are good in themselves and will in turn shelter
committed liberals who . . . and so on.

This means that we should do what we can to transform all states into
liberal states (starting at home); and since they will not remain liberal states
without a liberal political culture to support them, we should also do what
we can to ensure that there exist no political cultures except liberal ones. This
sets limits to enthusiasm for political–cultural pluralism. There may be arguments such as Mill sketched in *Liberty* for preserving enough individual conservatives, fundamentalists, bigots, flat-earthers, and whatever else to provide
liberals with the opposition they need to keep their argumentative muscles in
training, but it is implausible to think that we need bigoted regimes to keep
liberal regimes in good heart. An *advocatus diaboli* is one thing, but a dia-
bolical state another. Cultures in the broader sense occupy a middle ground,
and it is obvious enough that liberalism is not best promoted by
Kulturkampf; not only is cultural imperialism likely to produce backlashes of
a degree sufficient to undermine the entire project of dispersing liberalism
more widely, but in only a very few cases is the connection between cultural
allegiances and political arrangements so inflexible that it would be literally
inconceivable that a person would have to renounce all her or his cultural

attachments in order to operate political institutions that a liberal would find perfectly acceptable.

Even the most aggressive kind of consequentialist liberal—indeed, especially the most aggressive kind of consequentialist liberal—must be permitted the use of ordinary political intelligence. It is as obvious to him as it is to anyone else that nations should not in fact go around launching "civilisades" at their own whim and discretion. It does not follow that they ought not to—for instance—use their commercial muscle for the purpose, that they ought not to beam in propaganda, and find other means of influencing people and governments that do not threaten chaotic consequences. There will doubtless be states where propaganda in favor of free speech, a free press, the rule of law, frequent and uncorrupt elections, and universal suffrage will be received with the indifference that greets the double page advertisements for the thoughts of the Dear Leader that the North Korean government places in British broadsheet newspapers; but this will not always be true and not everyone will ignore it. The effort that the Soviet *bloc* used to put into jamming broadcasts from the West suggests that propaganda unnerves the ruling elite at least. Where regimes are simply intolerably brutal, the argument that states should not embark on military action for liberal purposes *ad hoc* is an argument against vigilante justice; but the proper replacement for vigilante justice is not to sit on our hands, but to establish a police force. Liberals who take themselves seriously should therefore try to form alliances up to and including something like a World Government, so as to have an orderly process of coercing the illiberal into decent behavior.[5]

This kind of consequentialist argument is familiar to everyone; welfare consequentialism is the operating theory of most democratic governments in domestic matters, and this is a more ideologically driven version of consequentialist reasoning in international relations. Of course, liberalism is also a theory of individual rights, and the explanation of what rights are and which rights we possess is not wholly easy in a consequentialist moral theory. Happily, that is not much to the point in this context; consequentialist liberalism is not a complete moral theory, and it does not have to commit itself on the ontological status of rights, even though it will mention few of them, and will be *insouciant* about the rights of states, though not those of individuals. The consequences to which consequentialist liberalism attends here may be regarded as already embedding liberal rights within them: that is, we are thinking in terms of maximizing the number of people and regimes who take the rights of individuals with sufficient seriousness, and who hold the values that encourage this. What is to the point in this context is that consequentialist arguments are open-ended in what they require; they may require action or inaction, coercive measures or fastidiously uncoercive measures. Since they require us to promote a goal, the two chief constraints are not to do what damages that goal, and not to do what threatens other sorts of damage worse than a failure to achieve that goal. A strong sense of practicality will focus on the danger that an action or string of actions will damage the value that it sets out to promote. Consequentialist liberalism is always vulnerable to the fear that an unduly aggressive liberalism will provoke such a backlash that the end will be worse than the beginning. Head-on assaults on religious orthodoxy, for instance, may provoke worse sectarianism; a softer and gentler approach may diminish it.

On the other hand, consequentialism is flexible. What no sane person would try to achieve by violence, we may achieve by education, not to mention nagging or seduction. There is no limit to the considerations that consequentialist arguments can properly take into account. And because there is a lot more to life than those parts of it that are best discussed in a liberal framework, it is unlikely that we shall think very often of doing anything that violates the ordinary taken-for-granted rights to life and security of person that governments exist to protect, unless the governments whose lives we mean to make difficult have themselves done something sufficiently dreadful—in particular, have begun to murder their own citizens on a scale that suggests that the casualties of war will be much lower than the victims of inaction, or have singled out some particular section of the poplulation for genocide. The aspirations of the "maximalist" will be different from those of the "minimalist," but the policies promoted by both are likely to be very similar; the difference between them is that the minimalist will think he has no right to contemplate doing what the maximalist thinks we have every right to contemplate, but usually every reason not to do.

Thus, on my analysis, Saudi Arabia is a society whose religious, social, and political structure ought to be pressured in a liberal direction by any liberal government that has relations with it. Elementary realism about oil supplies and military alliances urges on us greater caution than we might feel in many other cases, as does the anxiety that political instability might open the door to a still more reactionary government. Still, it is impossible to be a liberal and not think that the world is a worse place for the existence of the government of Saudi Arabia as presently constituted. There is undermining and undermining, however. One modest bit of undermining would be to make it clear on all occasions that its legal system, the repression of women, theocratic politics, and the rest of it are not charming local options but deeply offensive in the modern world. Another would be to make more fuss than we habitually do about the asymmetry of Saudi appeals for toleration in its favor while the regime practices wholesale religious intolerance itself. In the same way, the United States should be much sharper than it is about the Israeli government's readiness to act in brutal and unconsidered ways, and should not be deterred by the rhetoric of anti-anti-Semitism. Sometimes countries live up to their duty to nag their neighbors into good behavior; it was a good thing that the other countries in the European Union put pressure on the British government over the methods it used to police Northern Ireland and on more culturally touchy issues such as the right of homosexuals to serve in the armed services. Nagging is the obvious tactic when drastic measures would be foolish.

One familiar objection to this way of thinking rests on the claim that no state has a right to interfere in the internal workings of another, except in self-defense, and that every state has a right to the exercise of whatever rights inhere in its status as a sovereign state. That is exactly what this essay is arguing against. This essay rests on the assumption that sovereignty is not a fundamental right, and that states are not fundamental rights-holders. When we talk about rights, we should stick to the view that individuals have a good many negative human rights—particularly against ill-treatment by states, and one positive human right—the right to free action consistent with protecting the same right in others. States, on this view, have many moral obligations to their citizens, but none of their rights over their own citizens or against other states are human rights and all their rights are conventional; they are also vastly important, but they are important because well-observed conventions are important in international relations. That is not in contention. The point is that states are, in the crucial respects at issue here, quite unlike individuals. An individual may damage himself without violating the rights of others and is to be condemned for nothing more than imprudence, and not generally to be coerced out of his foolish behavior. A state cannot in the same way damage only itself; there is no "itself" to damage, only its citizens severally. The reality is thus that the wielders of coercive power are either neglecting their subjects' welfare or violating their subjects' rights, and probably both. The question then arises of what violations of individual rights legitimate intervention; I take it for granted that a bad enough neglect of the citizens' welfare will constitute a violation of their rights, since it will amount to starving them to death.

Common sense suggests that intervention—which will always be denounced as imperialism by some of those on the receiving end and elsewhere—is most justified and will secure the greatest support within the interventionist society when it is politically and culturally uncontroversial. This implies that minimalism will always triumph in practice. It also implies the priority of humanitarian intervention to prevent gross violations of a limited set of rights—to secure their rights against torture, massacre, and assault; intervention on these grounds is humanitarian rather than distinctively liberal, and overlaps with, but does not depend on a liberal view of the grounds for intervention. Arguments against intervening even on these grounds must be arguments of pure practicality, unless they are to hang on the thought that the answer to the question whether I am my brother's keeper is a firm no, that is, unless they are to rest on a very fierce realism. John Rawls and Michael Walzer have always argued that a philosophically defensible law of peoples will rule in humanitarian intervention but rule out intervention to establish a liberal state, and they have lately been joined by Michael Ignatieff. Their arguments have a good deal of force, but one *caveat* is that if we find that only more or less liberal states will reliably protect human rights in this minimalist sense, there may be little room to intervene on humanitarian grounds alone without embarking on regime creation too.

To see how far a consequentialist liberalism defence of intervention overlaps with arguments for intervention based on human rights, we can push this argument further. If the rights we think we should intervene to

protect include minimal political, social, and religious rights, we get the
beginnings of a liberal basis for intervention. By the same token, we get a
decreasingly strong case for thinking that the violation of rights *as such* jus-
tifies outside intervention. The question whether we should intervene thus
falls into a question of the consequences of so doing or not so doing. There
are two neatly opposed consequentialist arguments here. The first has just
been mentioned; if we intervene on humanitarian grounds and against the
will of the incumbent regime, as in Kosovo, we would wreck the local government by intervention and would have to rebuild it. We had better rebuild
it along liberal lines, at least in a minimalist sense. *How* minimalist is a good
question that is simply evaded here. Minimalism must at least mean that we
hardly look further than to the rule of law and fair elections; that is, we do
not consider such matters as whether local religious practices are scrupu-
lously egalitarian as between the sexes, what the state of the law on abortion
is, how egalitarian education is, and so on.

Conversely, anyone who doubted that a *liberal* as distinct from just any
sort of law-abiding and efficient government was required to avoid humanitarian disaster would argue that we ought to try to create a government that
is in tune with the local culture to the greatest extent possible without
risking the basic humanitarian goal. A government may be law abiding but
otherwise non-liberal; imagine a conservative Catholic government in nineteenth-century Europe, for instance. A political system built on a culture in
which there were intense social pressures for sexual and ethnic apartheid but
a fastidious attention to the rule of law and non-brutality in its enforcement
is not a completely incoherent notion, and if that turned out to be the most
stable regime to institute, the minimalist who is a humanitarian rather than
a liberal must follow that logic. The prudent liberal who accords priority
to humanitarian considerations but does not give up on the wider aim of
creating a more liberal world, will then think that the task is only half
accomplished and settle down to "nagging."

Finally, then, we come to the person whose views about the rights whose
violation licenses intervention—*if it works*—incorporate liberal maximalism,
and thinks a state violates our human rights in failing to provide us with adequate opportunities for self-expression, moral growth, and the development
of an autonomous personality. That view of human rights takes over the consequentialist's aim of maximizing the number of liberals in the world but
dresses it up in the language of rights. It thereby dilutes the notion of a right
by making it aspirational rather than peremptory. There is much to be said
against doing any such thing; first, because talk of rights is most appropriate
when we can show that what we are claiming is something that we can
demand and that the appropriate persons have a real duty and capacity to
provide it. *Non*-abuse—abstaining from torture, murder, robbery, and the
like is something we can demand from everyone, and governments in particular, and nobody has any excuse for not meeting their duty. Benefiting
from other people's positive contributions is always more difficult to think of
as a right, and positive contributions to the sustaining of a liberal culture

impossibly difficult. The Lubavicher Rebbe violates my civil rights if he clubs me about the head and utters racial insults while doing so; there is no argument to the effect that he violates my rights simply by holding illiberal views on many issues. If he violates anyone's rights, it is by the roundabout route of limiting the intellectual liberty of children who are brought up in narrowly orthodox households; and the analysis of parental rights on liberal principles is a subject this paper steers well clear of. Second, liberal goals are typically such as to be best promoted by argument, information, and non-coercive means, not only because that is the best way to make them stick, but because liberalism sets an intrinsic value on argument and information as routes to change. The maximalist will always be asking whether s/he is threatening the achievement of a liberal goal by adopting illiberal means. Arguments about rights are less adapted to such balancing; at best, they announce the conclusion of a balancing argument. Third, and however, even when the maximalist has agreed, first, that there is no human right to have our government create and sustain a liberal culture of the John Stuart Mill variety, second, that even if there were, it is absolutely unthinkable that it would be a right enforceable by outside intervention, and third and most important, that there is such a variety of liberalisms, culturally speaking, that one could not possibly have a right to the existence of any particular one, the maximalist will still want to say something like the following.[6]

Humanitarian considerations come first, both in the sense that they trump others in extreme cases, and that without a humane political environment, liberalism is an impossible project. Non self-defense based arguments for liberal imperialism will more often be broadly humanitarian than distinctively liberal. But it is a plausible hypothesis that what we ordinarily think of as liberal-democratic regimes are more reliable sustainers of humanitarian decency *for their own citizens* than any other; even if we think it sensible to adopt the usual international relations norms about coercive intervention— self-defence, first, humanitarianism, second, and nothing else to count in justification, whatever guidance it may provide once you've won—we should do what we can to spread political liberal democracy by all means short of war. And thirdly, it is a plausible conjecture that liberal democracies will last better if they are sustained by liberal sentiments in the political culture; what these are is matter for a different argument, but it seems plausible that they can come in many varieties and be grounded in many religious and cultural traditions. This is a thought on which it is hard to disagree with the Rawls of *Political Liberalism*. One can be a better Catholic siding with John XXIII than with Pius IX; one can be a liberal rather than an illiberal Catholic, just as one can be a liberal rather than an illiberal atheist. It implies that one should encourage liberal trends within whatever cultures we interact with, not that we should imagine ways to reduce them all to secular uniformity.

The minimalism that this paper takes seriously therefore is this: *in general*, we should in practice err on the side of caution; that is, we should *neither* contemplate the employment of the U.S. marines to maximise the number of good liberals in the world, nor even their employment to maximise the number of decent liberal regimes, however, the reason is not that we have no right to do it but that it just doesn't work. It does not work for too many reasons to spell out, but they begin with the difficulty of securing a consensus on policy in the intervening country, continue through the difficulty of knowing how to secure the goal we have in mind, and end with the justified resentment of the intervened upon who are treated like children in the interests of their own self-government. Even if it was more likely to work than it is, we still oughtn't to do it in other than very rare cases, since the liberal values of autonomy and self-reliance imply that it matters that people attend to their own liberation and that we respect their autonomy by allowing them to do so. The good liberal will adopt a liberalism that is consequentialist and unbounded in principle, but at anything beyond the level of agitprop s/he will be as cautious as anyone else. One further thing the good liberal will be conscious of is that the humanitarian motives that pull us into intervention— a distaste for bloodshed and cruelty—are exactly those that make war so disgusting in the eyes of liberals. It is all too easy to do more harm (defined in physical and emotional terms) than good, and that harm may very easily outweigh the value of the freedom brought into existence. One can irreparably damage liberal values by trying to implement them by force; and a creed that insists on the value of choice is the last creed that should be rammed down anyone's throat. Consequentialist liberals ought therefore to be tempted by imperialism—but under most conditions, they should resist.

NOTES

1. He represented his employment in the East India Company as a form of not too mindless drudgery that gave him enough spare time to write; seeing that by the end of his career, he occupied a position of much the same importance as a permanent secretary for the colonies, this was unduly dismissive.
2. Biographically, one perhaps should take into account Mill's view that as an irrevocable contract whose terms were profoundly disadvantageous to women, marriage was a form of slavery; objections to Mormons having lots of quasi-slaves on the part of people who thought it fine to have one each would have cut less ice with Mill than with most of his contemporaries.
3. Consider Mugabe's theft of the parliamentary and presidential elections in Zimbabwe; if it were possible to remove him "surgically"—as I do not for one moment imagine that it is—this would not be forcing Zimbabweans to be free in the sense at issue here, but analogous to the case of the hostage taker. It might, of course, be the case that if one asked the people of Zimbabwe whether they wished to be saved from Mugabe by outside invasion, they would answer that they did not; then they would be in the position of having been forced to be free. Analogies with Serbia and Iraq are there for the drawing.
4. Conor Cruse O'Brien, *The Long Affair: Jefferson and the French Revolution*
5. This is a fairly simple-minded thought in the following sense: the Europeans had such a police force in the form of NATO, and they did not use it. They could and should have used it, first to ensure that Croatia did not become an independent state before it had put in place measures to protect its Serb minority (whose memories of World War II do not incline them to trust a Croat

government), second to ensure that neither Croatia nor Serbia set about ruining the lives of the Muslim inhabitants of Bosnia-Herzogovina, and third to ensure that Serbia did not behave abominably in Kosovo. European governments could not bring themselves to act; the squeamishness that made them *feel* the horrors of what was happening also made them reluctant to get their own people killed. World War I having started in the streets of Sarajevo, so to speak, their fears were not absurd, but under the circumstances they were cowardly. So, having a police force is no use without the willingness to use it.

6. I hasten to point out that I go much further than Mill does in "A Few Words on Non-Intervention." Mill's line was that it was up to the oppressed to throw off their oppression and that outsiders ought not to do it for them; the British had no obligation to help the Hungarians turf out the Austrian monarchy. *However*, if the Austrian government were to rely on Russia to act in its old role as the gendarme of Europe, the British could come in on the side of the Hungarians. This is an argument to be handled with kid gloves; Gertrude Himmelfarb used to like it because she thought it justified the Vietnam War, but one can imagine the Bertrand Russell fan club thinking with equal warrant that it justified Russian intervention on the side of the Viet Cong—indigenous revolutionaries oppressed by a South Vietnamese government sustained by American power. Those of us who dislike talk of sovereignty don't much like treating nations as anything but a term of convenience; for us, the question is not *who* is involved but whether one can do something effective to establish a non-brutal, and preferably a non-brutal and minimally liberal, regime. The trouble with the brightest and the best is that they think it is easier to achieve than it is; hence the caution apparent in the paper.

CHAPTER 3

JEFFERSON, RIGHTS, AND THE PRIORITY OF FREEDOM OF CONSCIENCE

Jack N. Rakove

For many of his countrymen, Thomas Jefferson has become an unlikely authority to invoke on behalf of a modern conception of human rights. The embarrassing contradiction between his abstract commitment to equality and his self-indulgent life as the slave-owning master of Monticello has become too gross to bear. The strong likelihood that he fathered one or more enslaved children by Sally Hemings—his paramour, consort, or mistress, as well as half–sister-in-law—has only compounded the sin.[1] In the hands of Conor Cruise O'Brien, the distinguished diplomat and political commentator, Jefferson's glib rationalizations for the violence of the French Revolution foreshadow the naivete with which so many intellectuals condoned the murderous purges of Stalin or Mao, while his enthusiasm for states' rights places him closer to the rabidly anti-nationalist ideology of the so-called militia movement than to the new political order that most Americans have accepted since the 1930s.[2] And his famous endorsements of freedom of speech and the press notwithstanding, Jefferson's record as a civil libertarian has long been questioned.[3] Where Jefferson is concerned, the admonition to love the sinner but hate the sin may no longer apply.

Even so, one doubts that either public infatuation with or scholarly interest in Jefferson will abate soon. Books about the Sage of Monticello continue to pour from the presses, rivaled in quantity only by the Lincoln industry. More important, Jefferson remains a genuinely intriguing and compelling figure—not least for his contradictions or the elusive and evasive elements of his personality, but not only because of them, either. With Benjamin Franklin, whom he succeeded as the new republic's minister to the court of Versailles, Jefferson represents the American wing of the Enlightenment in the fullest sense of the term. So wide were his interests that one can imagine any of a number of words and topics within which to fill in the subtitular blank of *Jefferson and* _____. One such subtitle could easily be *Jefferson and Human Rights*—a topic that still awaits definitive treatment. For despite the vast literature devoted to both the Declaration of Independence and to particular facets of Jefferson's political and constitutional thinking, the subject has yet to be viewed in its entirety.

In this chapter, I propose to discuss three facets of Jefferson's thinking about rights: two somewhat concisely, and one at greater length. The concise discussions will address Jefferson's stature as a spokesman for the right of self-determination, and the distinctive and innovative aspects of his thinking about the nature and function of declarations of rights. The (modestly) lengthier section will be devoted to the one right that Jefferson and his close friend and political ally, James Madison, arguably regarded as the paradigmatic foundation of their essentially liberal conception of rights: freedom of conscience, which is, in turn, closely related to their strong commitment to the separation of church and state. In an age when theocracy is again acquiring some allure, when religiously rooted violence and persecution have become distressingly common, and when persistent efforts are afoot in the United States to lower the "high wall" of separation between church and state that Jefferson wanted to erect, there is something to be said for asking why the religion problem occupied so prominent a place in his political thinking, and whether his ideas have any relevance today.

JEFFERSON AND SELF-DETERMINATION

If we all know one thing about Jefferson, it is that he inserted the words "All men are created equal" into the Declaration of Independence—and the world has not been the same since. There were comparable statements of equality in other documents framed at the moment of American independence, such as the declarations of rights that accompanied many of the new constitutions of government adopted by the individual states. However, none of those other statements is as elegant as Jefferson's, and none enjoys the same authority. Only a scholar or his students would bother to read the state declarations of rights. But anyone with a middle-school education or up, in the United States or abroad, can understand and respond to the message of Jefferson's preamble to the Declaration. All of us being created equal, we are equally empowered to interpret this text for ourselves, and Jefferson (like Thomas Paine) helps us by writing with a directness and clarity of expression that we rarely find in eighteenth-century writers.

That empowerment can lead to results that scholars find problematic. Take the single word immediately preceding the equality statement: "We hold these truths to be *self-evident*." To a modern reader, the colloquial self-evident meaning of "self-evident" is something like "obvious." In eighteenth-century usage, however, "self-evident" was more akin to "axiomatic," still suggesting, perhaps, that the truths being announced are generally accessible, yet requiring a measure of reflection (in the way, for example, that our mind's eye has to imagine why parallel lines never intersect).[4]

The greater error into which individual readers can easily fall is to assume that when Jefferson spoke of all men bring created equal, he was primarily describing an equality of individuals rather than an equality of peoples. That interpretation indeed became increasingly more appealing with each passing decade, as one group of outsiders after another sought to harness its rhetorical power to their particular causes. If all men (or persons) are created equal, and as a result are "endowed with certain inalienable rights" to "life, liberty, and the pursuit of happiness," then all discriminations among different classes of the population must ultimately appear problematic and arbitrary (or at the very least require earnest and rational justification). Why should unpropertied males, or unenslaved African Americans, or propertied women be denied the same rights as propertied white males? Why should age, class, creed, race, country of origin, or gender be allowed to restrict our equal rights of individual opportunity?

But in writing the preamble to the Declaration, Jefferson was seeking neither to strike a blow for the equality of individuals, nor to erase the countless social differences that the law sometimes created and often sustained. The primary form of equality that the preamble asserts is an equality among peoples, defined as self-governing communities. The situation confronting the Americans in 1776 did not presuppose a return to a state of nature, where individuals possess equal natural rights, but only the condition known as "dissolution of government" created by the unjust and violent acts of the British empire. The right being asserted is that of "one people to dissolve the political bands which have connected them with another," and in doing so, to replace one form of government with another.

This equality among self-governing peoples does not require Jefferson to assert the perfect legal equality of individuals within the communities so constituted. Nor does it require him to suggest or conclude that all peoples, once constituted, have exactly the same endowments. One of the most striking (if brief) echoes of the Declaration found in Jefferson's later writings occurs in the opening passage of the (in)famous Query XIV of the *Notes on the State of Virginia*. Describing his proposal to emancipate and colonize "all slaves born after passing the act," Jefferson says that Virginia should "declare them a free and independent people and extend to them our alliance and protection, till they shall have acquired strength."[5] The Declaration of 1776 had closed by asserting that the Americans were now "free and independent states," with "full power to levy war, conclude peace, contract alliances"— that is, to seek alliance and a measure of protection from others, until their own independence was secured. But this echo of the Declaration is quickly followed by Jefferson's labored attempt to justify the colonization of African Americans on the racialist grounds that in their natural faculties they appear permanently inferior to whites, and therefore incapable of exercising an equality of individual rights with their former masters.

This is the Jefferson about whom Americans now feel so uncertain and nervous, skeptical and critical. It would not be difficult to condemn his proposal for the involuntary expatriation of the freedmen, nominally undertaken as an act of national liberation that will create a new people, as a form of ethnic cleansing. Jefferson stumbles over this difficulty in another famous passage of the *Notes*, Query XVIII, when he describes how slavery corrupts the republican morals of the master class. "The man must be a prodigy who can retain his manners and morals undepraved by such circumstances," Jefferson observes, and then proceeds to an equally telling observation:

> And with what execration should the statesman be loaded, who permits one half the citizens thus to trample the rights of the other, transform those into despots, and these into enemies, destroy the morals of one part, and the amor patriae of the other. For if a slave can have a country in this world, it must be any other in preference to that in which he is born to live and labour for another: in which he must lock up the faculties of his nature, contribute as far as depends on his individual endeavours to the evanishment of the human race, or entail his own miserable condition on the endless generations proceeding from him.[6]

This passage is as remarkable as it is troubling. On the one hand, Jefferson recognizes slaves as citizens *in potentia* who have lost their natural right to liberty and with it the capacity to feel the natural love of country that binds Jefferson to his native state. On the other, Jefferson admits no possibility that emancipation and the conferral of rights can create a love of country among a previously exploited population. Instead he assumes, as he had previously explained in Query XIV, that "Deep rooted prejudices entertained by the whites; ten thousand recollections, by the blacks, of the injuries they have sustained; new provocations; the real distinctions which nature has made, will divide us into parties, and produce convulsions which will probably never end but in extermination of the one or the other race." No room here for the leavening power of truth and reconciliation commissions: the permanence of historical memory and "the real distinctions" of race create a doomed legacy.

It does not help Jefferson's case that he could never identify the territory to which the freedmen could be colonized, or measure the nature of the debt that their former masters would have to pay. The country that is to support this involuntary emigration is not, after all, the United States but his own Virginia, and it is difficult to imagine—given his state's likely low level of public investment and his countrymen's aversion to taxation—that the aid to be extended would amount to much, especially if Virginia would at the same time have to set about to recruit "an equal number of white inhabitants" to supply the place of the exiled blacks. The bitterest element of Jefferson's retirement arose from his frustrations in getting a penurious legislature to provide adequate funds for the University he founded at Charlottesville, much less the system of common schools he hoped to establish; what kind of compassionate altruism could he have expected the master class to bestow on its former property? Jefferson's sense that Virginia was his true country, and his emphasis on racial identity as the fundamental criterion of membership, look suspiciously like the romantic nationalisms that would flourish in the nineteenth century and again in our own times. Or perhaps we should not be so harsh. Why not justify and celebrate Jefferson as the prophet of national self-determination, asserting the rights of a colonial people on a basis of political equality with their former imperial masters?

JEFFERSON AND THE PROBLEM OF DECLARING RIGHTS

If Jefferson could have had his way back in 1776, his historical reputation would not depend on his authorship of the Declaration of Independence. At the time he was stuck in Philadelphia, drafting the Declaration, he would have preferred to return to his old college town of Williamsburg, to join in writing the new constitution of state government. The formation of these new governments "is the whole object of the present controversy," he observed rather boldly, if inaccurately, in May 1776.[7] As fate had it, his work at the Congress proved more important and monumental. The draft constitutions he prepared in 1776, and again in 1783, are interesting primarily as guides to Jefferson's political thinking, rather than as reflections of the process of constitution-making that was unfolding within the states while he was drawing up his indictment of George III.

In fact, Jefferson's notion of constitution-writing is highly revealing precisely because, on certain key issues, his thought was indeed distinctive. This is especially the case in his approach to the problem of declaring rights—that is, deciding what kinds of declarations of rights should accompany the new frames of government that the provincial conventions were drafting as the colonies moved toward independence.

Today, of course, we are habituated to the idea that written declarations of rights provide a fundamental source of legal, constitutional, and moral authority to which aggrieved citizens and groups can resort for protection. Americans think of bills of rights as judicially enforceable protections, and much of the modern movement for international human rights rests on similar assumptions. To be effective, international statements of human rights must somehow become enforceable against the intervening claims of national sovereignty. Such statements also need to be specific. It will not do to fashion the claims of rights in broad platitudes, and hope that governments will henceforth do the right thing. They must also identify particular activities or aspects of the human personality deserving protection against the arbitrary power of either the state or the dominant groups within society.

When Americans first drafted bills of rights in 1776, however, no such simple conception of their function was available. The colonists had some antecedents to draw upon: the parliamentary Bill of Rights of 1689, or the catalogs of "liberties and privileges" that sometimes were adopted by the colonial legislatures. But these were hardly constitutional documents in the modern sense. Indeed, how could they be, when the modern definition of a constitution as written, supreme, fundamental law was exactly what the American revolutionaries were still in the process of inventing?

The interpretation of these first bills of rights accordingly poses something of a puzzle. For one thing, these declarations accompanied the first constitutions but were not, strictly speaking, a part of them. Many of the articles included in these declarations merely affirmed a general principle of government without specifying how that principle was to be implemented. When a particular right was identified, the authors preferred to say that it "ought" to be protected rather than it "shall" be protected. No particular institution of government was made responsible for the enforcement of these rights. Indeed, a good case can be made that the first declarations were addressed primarily to the citizens, so that they would know or recall what their rights were, and be able to judge the behavior of their rulers accordingly. Bills of rights were essentially political statements. They were appropriately issued at a special moment in history when a people was reconstituting itself as a self-governing community, as a statement of purpose and principle.

Jefferson, however, had fashioned a different and more advanced conception of the relation between declarations of rights and formal constitutions of government. In the draft constitution that he sent from Philadelphia to correspondents in Virginia, the fourth and concluding article took the form of a statement of "Rights Private and Public."[8] In three respects, this passage stands as one of the more intriguing texts of American constitutional history; yet its significance has been largely neglected by those who see Jefferson's concurrent authorship of the Declaration of Independence as his paradigmatic statement on rights.

The first point of significance is a formal one: Jefferson was the first American to appreciate the value of incorporating statements of rights directly into the body of a written constitution, rather than cast them as free-standing documents of uncertain constitutional authority or legal effect. In the modern parlance, he was an advocate for the *entrenchment* of rights, henceforth to be understood not as general principles or guidelines that hopefully would be followed, but rather as explicit constitutional commitments. Given that Jefferson also numbered among the first Americans to appreciate the fundamental distinction between a written constitution as supreme law and ordinary legislation, this commitment was a substantial one.

Second, Jefferson's statement of rights is intriguing because it generally omitted most of the standard articles found in the declarations of rights the states were then adopting. In Gordon Wood's phrase, those declarations typically offered "a jarring but exciting combination of ringing declarations of principle with a motley collection of common law procedures."[9] *Motley* is too disparaging a term to apply to the procedural safeguards that protected citizens and subjects against the coercive power of the state, but Wood's characterization otherwise holds. Jefferson valued those procedures and principles no less than his countrymen, but he did not think it essential to place them in the text of his constitution.

This leads in turn to the third and most revealing difference between Jefferson's approach to rights, ca. 1776, and that of his contemporaries. For on the positive side of the allocation of rights, he envisioned the state providing 50 acres of land "in full and absolute dominion" to "every person of full age neither owning nor having owned" the same quantity of land. That was double the amount of land required (in Article I of Jefferson's draft constitution) to exercise the suffrage or be eligible for election to any office, meaning that Jefferson contemplated not only the elimination of any class of tenant farmers (of whom there had been significant numbers in the colony's history), but also the broad distribution of political rights to the adult male population. Jefferson does not make clear whether the appropriation of land shall go to women as well as men—he says "every person" in the relevant article, but refers to "all male persons of full age and sane mind" when discussing political rights—but it is noteworthy that one clause of his fourth article stated that "Descents shall go according to the laws of gavelkind [equal partible inheritance], save only that females shall have equal rights with males." Jefferson also would have used his article on rights on behalf of the racial minorities on whose land and labor white society relied. Indian lands were to be purchased from their "native proprietors . . . on behalf of the public," with every purchase taking place under the specific authorization of the legislature. Further, "No person hereafter coming into this country shall be held within the same in slavery under any pretext whatever." Only after advancing these novel proposals did Jefferson mention several of the other rights that other Americans generally expected to see included in a declaration: "full and free liberty of religious opinion"; a right of a freeman to use arms "within his own lands or tenements"; a prohibition on standing armies, save in wartime; and freedom of the press, except for private injury.

What is the theory of rights that this document expresses? It rests not on the identification of basic civil liberties or the declaration of fundamental republican principles, but rather on a conception of universal citizenship in which all free members of society would receive the endowment required for effective participation in *res publica*. It assumes, moreover, that this is no mere ideal to be attained at some indefinite point, but an actual social policy expressed as a constitutional commitment. Its approach to the suffrage stands in interesting contrast to the position that John Adams—Jefferson's colleague on the committee drafting the Declaration—was taking at exactly the same time. In his own fascinating meditation on the suffrage, Adams warned that the moment of revolution was not the best time to ask whether artisans or women were entitled to vote. It would be "dangerous to open So fruitfull a Source of Controversy and Altercation, as would be opened by attempting to alter the Qualifications of Voters," Adams observed in May 1776. "There will be no end to it." Rather than confront the question directly, as a matter of principle, it was better to make sure that "the Acquisition of Land [is] easy to every Member of Society," in which case "the multitude" would always be capable of protecting its "Liberty, Virtue, and Interest" against threats of domination.[10]

Adams and Jefferson thus agreed that an equitable distribution of property was essential to the protection of rights; where they differed was whether that end should be expressed as a matter of constitutional commitment. Adams was no less an enthusiast for republican government than

Jefferson; his own pamphlet, *Thoughts on Government*, exerted far more influence over the constitutional deliberations of 1776 than anything Jefferson wrote. Yet where Adams had the good sense to suggest that some issues should be deferred or resolved discreetly, Jefferson's draft constitution would presumably require imminent and explicit implementation.

In his convictions, Jefferson appears to have been one of those revolutionaries inclined to believe that the expedient needs of the struggle should not claim priority over the just ends for which it is being waged. Others could argue that there would be time enough to think about justice, the optimal design of constitutions, and the enforcement of rights once independence was secured. Jefferson preferred to err on the other side of the question. That was the meaning of his original comment that the proper design of new constitutions of government, and not merely independence, was "the whole object of the present controversy." That was also why his major achievement after 1776 was to throw himself wholeheartedly into a project to revise the entire Virginia statutory code, and this during a period (1777–79) when the outcome of the war remained uncertain, and when there was no chance that the Virginia legislature could possibly consider the revised code of laws he was studiously drafting. Jefferson's detractors can plausibly object that, after reluctantly writing the Declaration, his actual service to the revolution was hardly noteworthy. Had not his later career and fame as party leader and president guaranteed his stature and reputation, he might be recalled as an interesting embodiment of the American Enlightenment but a decidedly second-rank leader.

Yet from the modern vantage point of human rights, there is much to be said for a revolutionary who resists relegating the cause of constitutionalism and rights to a subordinate and distant position behind the original struggle for victory and security. It is all too easy and common, after all, for victorious revolutionaries to equate their own authority and charisma with the revolutionary cause itself, to the dilution and sacrifice of the principles for which the struggle was waged or the true ends for which it ought to have been fought. Or it may be all too common for the people at large to neglect the public good that had united them in a spirit of sacrifice and turn instead to the pursuit of their own narrow interests. "It can never be too often repeated," Jefferson observed in the *Notes on the State of Virginia*, "that the time for fixing every essential right on a legal basis is while our rulers are honest, and our people united. From the conclusion of this war, we shall be going down hill."[11] Fittingly, the context of this remark was Jefferson's discussion of freedom of conscience.

THE PRIORITY OF FREEDOM OF CONSCIENCE

Of the 126 bills that Jefferson prepared in his revised code of Virginia statutes, the one that has attracted the greatest notice, and received the greatest acclaim, is Bill No. 82, "For establishing religious freedom."[12] Jefferson drafted the bill in the late 1770s, and his ally, Madison, finally secured its enactment in 1786. To the untutored reader, or even a modern jurist, the Statute for Religious Freedom is a curious document. Most of its text—the lengthy preamble consisting of a single sentence of Faulknerian proportions—provides a philosophical and boldly rhetorical justification for the measure. The legally efficacious core of the Act is a bare seventh the length of the preamble, and provides that no one can be "compelled to frequent or support any religious worship," or to suffer "on account of his religious opinions or belief"; rather, all "shall be free to profess, and by argument to maintain, their opinions in matters of religion," with no effect on "their civil capacities" (that is, their rights). A concluding paragraph affirms that this statute, like any other, is subject to repeal, but that such a repeal "will be an infringement of [the] natural right" that the act merely recognizes (or literally, declares).[13]

Why is this statute so important, and how might its interpretation and application suggest a critical link between the concerns of this representative expression of the eighteenth-century Enlightenment, and the post-Enlightenment regime of human rights in the twenty-first century?

We can start to answer these questions by recognizing that the religion question no longer confronted eighteenth-century Anglo-American thinkers with the bloody intensity with which it had shattered the peace of sixteenth- and seventeenth-century Europe. The killing of heretics, the burning of martyrs, the slaughter of innocent members of dissenting sects, the waging of civil and international wars on religious lines: all were now part of the historical past. In the Old World, all kinds of questions remained about how far toleration should extend, and lawful discriminations of numerous kinds still applied against dissenters and infidels. In the New World, however, where public authority was weak to begin with, there were few meaningful forms of discrimination to apply, while the religious passions which had burnt so fiercely (at least in Puritan New England) had clearly abated.

Thus the godly commonwealth of Massachusetts, for example, had lost the legal authority to execute the occasional Quakers who refused the colony's warnings to stay away, and even had it retained that power, eighteenth-century sensibilities would have prevented its exercise. Modest attempts to maintain religious uniformity were made in the wake of the religious revivals of the 1740s and 1750s, which imposed serious strains within denominations, but never threatened violence. The instances of sectarian animosity recorded by the Anglican missionary, Charles Woodmason, during his travels in the Carolina backcountry are the stuff of low comedy—Scots-Irish Presbyterians, for example, disrupting his service by unleashing a pack of dogs, and he returning one to its owner with the information that he had made a convert.[14] James Madison, a somewhat stern young philosophe just returned to rustic Virginia from his collegiate studies at Princeton, could burn with righteous anger in 1774 over the brief jailing and fining of a half dozen itinerant Baptists, charged with preaching without a license from the civil authorities. However, what agitated Madison was less the overt persecution the Baptists were suffering, but its sheer anachronism. Enforcement of religious conformity was an embarrassing relic of another age, but hardly a major source of oppression.[15] For enlightened thinkers like Madison and Jefferson, the history of religious persecution had become the equivalent of what the history of racial discrimination is for us: the most compelling illustration of what the arbitrary exercise of coercive power and the denial of fundamental rights could mean in practice. (The irony, of course, is that we expect them to have shared the same sense of moral revulsion over racial prejudice that we do—not realizing that such attitudes remained a more familiar and acceptable element of their world.)

To say that the repudiation of religious intolerance had become enlightened orthodoxy, however, is not to explain its deeper significance in their time or for ours. That significance rests on at least three major propositions, each of which can be associated with Jefferson and Madison. First, freedom of conscience—that is, the right to choose what to believe or not to believe—should be regarded not only as a fundamental natural right, but also as the quintessential liberal right that recognizes the capacity of individuals to preserve a realm of privacy where the authority of the state cannot intrude. Second, by favoring the principle of disestablishment as the best means to secure freedom of conscience, the Virginians were also clarifying the basic concept of limited government that we associate with the idea that government consists of a limited delegation of power from the community at large. In this sense, religion represented the first major form of behavior that could be safely removed from the regulatory purview of the state—in effect, privatized. Third, most important, and most controversially, this solution to the religion question carries a universalist allure that should seem particularly attractive today, at a time when religious conflicts seem to have recovered their frightening capacity to inflict massive harm.

Jefferson and Madison first met in the fall of 1776, when both served on the Virginia legislature's committee on religion, which considered how far the state should go in recognizing freedom of conscience and moving toward disestablishing the Anglican church. The answer to these questions, to their way of thinking, was: not far enough. Madison had already gained the first success of his long legislative career back in May, securing an amendment broadening the principle of religious freedom in the Virginia Declaration of Rights from a mere statement of toleration to a recognition that "all men are equally entitled to enjoy the free exercise of religion, according to the dictates of conscience." Jefferson, too, was not content with a policy of mere toleration. In the reading notes he prepared during this period, Jefferson included this revealing comment on John Locke's *Letter Concerning Toleration*. "It was a great thing to go so far (as he [Locke] himself sais of the parl[iament] who framed the act of toler[atio]n" of 1689, Jefferson wrote; "but where he stopped short, we may go on."[16]

What would it mean to go beyond the position that Locke had espoused a century earlier? Locke's approach to the religion problem had both philosophical and political dimensions. On the first of these, the Virginians were essentially derivative and unoriginal. It was in the second area—the realm of what they could imagine as politically feasible—that their position found its radical character and implications.

The philosophical foundation for toleration rests on the inner nature of religious belief and conviction, which can neither be manufactured by coercion nor accurately monitored by external authority. Although operating within a framework that remained decidedly Christian and Protestant, Locke also conceded the impossibility of knowing the best path to the otherworldly salvation which is the Christian's true reward, and hence was prepared to concede a general right to everyone "to perform openly that form of worship by which he or she seeks salvation." But this right (or privilege) was not universal. Locke's well-known exceptions from the toleration of the state included Catholics or atheists, while others were left "in a vague limbo without any clear status or guaranteed freedom."[17] These were groups whose loyalty and behavior were questionable: Catholics because they owed allegiance to an alien power, atheists because the absence of fear of punishment in another world made them unreliable subjects in this. No state could sensibly renounce its authority to regulate the conduct and condition of those of doubtful conviction. Toleration was thus a privilege to be conceded, but it very existence supposed that the state retained a power it could not safely renounce.

When Jefferson imagined that the Americans might find a way to "go on" beyond mere toleration, he took for granted the psychological dimensions of religious belief that Locke had laid bare. True religion was a matter of inner belief, which could safely manifest itself in any outward performance of ritual or conduct so long as no harm was done to another—or presumably so long as the actions of one sect did not offensively target the beliefs of others.[18] But on matters supernatural, all belief was ultimately a matter of opinion, and on the matter of religious opinion, Jefferson was prepared to let a thousand doctrines—even a thousand deities—bloom. Jefferson put the key point in the most direct language possible.

> The legitimate powers of government extend to such acts only as are injurious to others. But it does me no injury for my neighbour to say there are twenty gods, or no god. It neither picks my pocket nor breaks my leg.

Jefferson held out the hope that "free enquiry" in the realm of religion would be a source of progress, of a sort, but the improvement to come would seemingly take the form of exposing sources of religious credulity and tyranny for the errors they were. It was another matter entirely to think that a regime of perfect toleration or the widest application of the free-exercise principle would lead to a "uniformity" of belief or conviction. "Difference of opinion is advantageous in religion," Jefferson opined, because the rival sects act as a "Censor morum" over each other. But "Is uniformity attainable?" he then asked. "Millions of innocent men, women, and children, since the introduction of Christianity, have been burnt, tortured, fined, imprisoned; yet we have not advanced one inch toward uniformity." On a planet of a billion people, there were "probably a thousand different systems of religion," and none—not even Christianity—had a realistic hope of bringing itself to a complete consensus, much less persuading others of its all-conquering truth.[19]

Jefferson had indeed come a long way from Locke's underlying commitment to a Protestant Christianity and his potent reservations about tolerating Catholics and atheists, much less conceding, as a matter of civil policy, their natural right to believe as they wish. To be sure, Jefferson had his prudent moments, too. The Statute for Religious Freedom evoked the authority of an "Almighty God" who was "the holy author of our religion"—presumably not a reference to Zeus or Krishna, but a politic nod to the God of the Old and New Testaments and possibly the Son who appears in the latter. Like Locke, however, Jefferson retained a religious sensibility. This was a theology that he was espousing, not an anti-theology, and Jefferson was serious enough in its pursuit to produce his own rendition of the Christian Gospel, purged of all its miraculous trappings to distill the original and sublime message of Jesus, who appears not as the messiah but as a perfect prophet.[20] Jefferson had ceased to be a Christian, in the orthodox sense, but he remained a radical Protestant, willing to trust the laity in their pursuit of faith.

This recognition of a private, subjective right of judgment in the realm of conscience should be acknowledged as the radical departure in thinking about rights more generally that it represents. Locke's position had arguably been the more sensible one. Toleration was wise as a matter of practice, and philosophically justified on a theory of belief that recognized the limits of the magistrate's power either to discover spiritual truth or to probe the inner convictions of subjects. But no state could afford to renounce its power to define the outer limits of permissible belief, for the simple reason that some forms of religious dissent, if left unregulated, posed too great a threat to the peace, stability, and security of society. Catholics and atheists could always exercise their *natural* right of conscience to believe what they wished, but their *civil* right to do so unregulated by the state remained problematic. Jefferson, by contrast, was content to let belief go unregulated, confident that the competition among sects and the progress of reason would prevent the dangerous enthusiasms and persecutions of the past from recurring in America.

It is this radical trust in the capacity of ordinary individuals to form their own spiritual convictions, and of sectarian rivalry to take the form of a marketplace of competing religious beliefs, that reveals the radical implications of Jefferson's (and Madison's) position for human rights more generally. Freedom of conscience is the paradigmatic or essential individual right, because it is the one that makes the strongest presumption of an individual's sovereign capacity, in the inner recesses of his or her mind (and soul), to decide which religious ideas correspond most strongly with our own opinionated convictions. This distinguishes it from most of the other civil rights that we wrongly tend to regard as similarly absolute in nature. In fact, most of the other rights recognized in American-style bills of rights are essentially procedural. They assume, that is, that government does have a power to act, or to regulate, but when it chooses to do so, it must conform to some set of fixed rules and principles that respect our interest in preserving the liberty and autonomy of the citizen. Our homes may be our castles, but any deputy armed with a search warrant can rummage through our drawers and closets. Americans are far less constrained, too, in the marketplace of religious beliefs than in the free market of investment, commerce, and labor that we love to celebrate, much less the market for competing political ideas. So, too, the right to choose (or reject) religious leaders historically took precedence over the right to choose our elected representatives.

Of course, not all markets are perfect, and Jefferson spent his declining years worrying over how well the commerce in religious opinions was operating. Freedom of conscience and disestablishment should have worked to lessen the hold of the more enthusiastic and revelation-based forms of religion, he hoped, while making a reasoned deism and common notions of morality the defining elements of American religiosity. It would be a religiosity founded on the simple teachings of Jesus—as opposed to the doctrines propounded by most churches in his name since, with Calvinists now ranking high on his list of reprobates—lodged in a nation where every "*young man* now living," he wrote in 1822, would "die an Unitarian"(no longer believing in the divinity of Jesus).[21] But the aging statesman now lived in a country (and state) where Baptists and Methodists were flourishing, plunging the converted into rivers everywhere and holding frenzied revivals where the interior expression of religious conviction took the external form of holy rolling and frenzied weeping, where the new revelations of Joseph Smith were on the horizon, and where his own plans for a secular University were harassed at every step by religious enthusiasts worried about the wages of impiety.

It was James Madison, I think, who better grasped the implications of the emerging American religious system. His own religious opinions remained militantly private, and hence obscured from the historian's gaze. In the absence of hard evidence, one can accordingly ascribe to Madison a whole gamut of plausible positions, ranging from a residual Calvinism nurtured in his years at that Presbyterian college in Princeton to a philosophical deism from which religious faith of any kind was absent. Madison seems not to have cared what forms American religiosity was taking. In his long retirement, he had the satisfaction of knowing that experience had falsified the gloomy predictions that had once met his and Jefferson's brief for disestablishment back in the 1780s. Then opponents had warned that religion would decay without the continued support of the state; but the opposite result had obtained. Religion had never been more vibrant or spirited. It did not matter whether or not the rivalry of the sects and denominations was fostering a more refined form of religiosity, as Jefferson had hoped. What did matter was simply that disestablishment promoted a healthy competition demonstrating that the natural right to freedom of conscience could be fully indulged as a civil right as well, with no harm to religion itself, and positive benefit to the state. Jefferson might still worry that Bible-reading Americans would arrive at all kinds of strange and discordant conclusions; Madison was relieved to think that the resulting diversity of religious opinion would help maintain the "multiplicity of interests" that provided the cure for the mischiefs of faction he had hypothesized in *Federalist* 10.

IMPLICATIONS?

Historians make two not entirely consistent claims on behalf of their knowledge of the past and its salience to the present. One is that we cannot fully comprehend the world in which we live without understanding how it emerged from the past. The other is that the past is (was) a different place—a foreign country, we like to say—and we should be careful to distinguish its underlying differentness from the present. The historian's best lesson from the past may be to avoid drawing any lessons at all—to look for distinctions rather than similarities—while still striving for an awareness of the connectedness of past and present.

Even so, to someone who remains frightened, not to say depressed, by the strange persistence of religious modes of reasoning and budding theocracies in an ostensibly scientific and secular age, there is something deeply appealing about the solution to the religion problem that Jefferson and Madison pioneered. In the context of the distinctive American experience, its significance seems fivefold. First, its emphasis on our natural and inescapable capacity to assess the truth claims of religion laid the foundation for the modern liberal conception of the citizen as an individual subject-actor possessed of a realm of private belief where the authority of the state cannot intrude. Second, the privatization of religious belief and behavior identified the first significant realm of human activity that the state could no longer regulate, and therefore exemplified the fundamental constitutional principle of limited government: that the powers exercised by the state require a positive act of delegation. Third, once these conditions were established, religiosity in general would thrive, not wither. In the marketplace of beliefs and interpretations generated by the underlying Protestantism of American culture, sects, denominations, and whole faiths would compete for adherents, and conditions of monopoly or oligopoly would be impossible to maintain. Fourth, this peaceful competition would engender attitudes of mutual toleration and respect—if not always toward the irreligious or openly profane, then at least toward the adherents of other religious opinions.[22] Fifth, while the ensuing religiosity of the population in general would mean that many participants in public life would act on the basis of their religious convictions, the diversity of faiths and denominations would make it difficult for these groups to coalesce in any sustained way, especially (if one follows the Madisonian calculus) at the national level.

Can these "lessons" or developments be generalized to other cultures, or are they best viewed as another manifestation of American exceptionalism?

One obvious objection or caution is that the success of the ideology owes less to principle than to the underlying and primarily demographic sources of diversity within the population. A second, more telling objection is that the approach of Jefferson and Madison depends upon a fundamentally Protestant understanding of religion itself. That is, the true or proper expression of religiosity is a matter of inner conviction and the capacity to perform innocuous rituals or modes of worship consistent with one's *individual* belief about the nature of salvation in a world to come. But this is a form of religiosity which does not insist that all of life must be lived in conformity with the dictates of religious law, and it cannot tolerate the idea that religious authority is a sufficient warrant for secular law. It may acknowledge that the reigning legal system will reflect or be informed by religious values, in some general, diffuse way, but it will not insist, as the first settlers of Massachusetts Bay once did, that the code of civil law be consistent with or derived from "Moses, His Judicials." Nor, of course, can a definition of religious experience as primarily internal and individual in nature be reconciled with a view of religion as a communal expression of one true faith's identity and supremacy.

Still, I like to think, no doubt naively, that Jefferson and Madison—building on Locke, but unknowingly verging on Spinoza[23]—basically got it right, and the whole world would be a better, safer place if their principles somehow gained universal acceptance. For in the end, the basic insight on which the principle of freedom of conscience rests is the recognition that the religious beliefs of *all* faiths are necessarily a matter of inner opinion, and that opinion cannot be coerced or accurately measured or ascertained. Whatever the source of divine revelation or prophecy, none of us knows any more or less than anyone else about the supernatural or the spirit or the world to come. It is all a matter of opinion, as Jefferson understood even as he worked out his own enlightened private theology.

NOTES

1. For my own reflections on this subject, see Jack Rakove, "Our Jefferson," in Peter S. Onuf and Jan Lewis, eds., *Sally Hemings and Thomas Jefferson: History, Memory, and Civic Culture* (Charlottesville: University of Virginia Press, 1999), 210–35. The current essay draws upon ideas expressed there and in idem., "Once More into the Breach: Reflections on Jefferson, Madison, and the Religion Problem," in Diane Ravitch and Joseph Viteritti, eds., *Making Good Citizens: Education and Civic Society* (New Haven: Yale University Press, 2001), 233–62.

2. Conor Cruise O'Brien, *The Long Affair: Thomas Jefferson and the French Revolution, 1785–1800* (Chicago and London: University of Chicago Press, 1996).

3. Leonard Levy, *Jefferson and Civil Liberties: The Darker Side* (Cambridge: Harvard University Press, 1963).

4. Morton White, *Philosophy*, The Federalist, *and the Constitution* (New York: Oxford University Press, 1987), 208 ff.

5. Merrill Peterson, ed., *Thomas Jefferson: Writings* (New York: Library of America, 1984), 264. It is important to recall that Jefferson himself notes that this proposal was drafted as an amendment to another bill concerning slavery, but not formally included with it or introduced.

6. Ibid., 288–89.

7. Jefferson to Thomas Nelson, May 16, 1776, Julian P. Boyd, ed., *The Papers of Thomas Jefferson* (Princeton: Princeton University Press, 1950–), I, 292.

8. Ibid., 362–63.

9. Gordon S. Wood, *The Creation of the American Republic, 1776–1787* (Chapel Hill: University of North Carolina Press, 1969), 271.

10. Adams to James Sullivan, May 26, 1776, in Robert Taylor, ed., *Papers of John Adams* (Cambridge: Harvard University Press, 1977–), I, 208–12.

11. Peterson, *Jefferson: Writings*, 287.

12. It is one of the few legislative enactments in American history to have an entire volume of scholarly essays devoted to its assessment. Merrill D. Peterson and Robert C. Vaughan, eds., *The Virginia Statute for Religious Freedom: Its Evolution and Consequences in American History* (New York and Cambridge, UK: Cambridge University Press, 1988).

13. Peterson, *Jefferson: Writings*, 346–48.

14. Richard J. Hooker, ed., *The Carolina Backcountry on the Eve of the Revolution: The Journal and Other Writings of Charles Woodmason, Anglican Itinerant* (Chapel Hill: University of North Carolina Press, 1953), 45–46.

15. Madison to William Bradford, Jan. 24, 1774, in Jack N. Rakove, ed., *James Madison: Writings* (New York: Library of America, 1999), 7.

16. Boyd, *Papers of Jefferson*, I, 548.

17. Jonathan L. Israel, "Spinoza, Locke, and the Enlightenment Battle for Toleration," in Ole Peter Grell and Roy Porter, eds., *Toleration in Enlightenment Europe* (Cambridge, UK: Cambridge University Press, 2000), 103–04.

18. Hence Jefferson's revised code of laws could plausibly include a bill for the punishment of sabbath-breakers. As president, however, Jefferson insisted on the Sunday delivery of the mail, because respecting the Christian sabbath in this way would amount to an establishment of the religious beliefs of one group in derogation of those of others.

19. Query XVII, *Notes on Virginia*, in Peterson, *Jefferson: Writings*, 285–86.

20. See Thomas J. Buckley, S.J., "The Political Theology of Thomas Jefferson," in Peterson and Vaughan, eds., *Virginia Statute for Religious Freedom*, 75–108.

21. Jefferson to Benjamin Waterhouse, June 26, 1822, *Jefferson: Writings*, 1458–459.

22. One of the curious features of American culture is that both the faithful and the skeptical tend to believe that their convictions are not given adequate respect. The former think that their beliefs are scorned by dominant intellectual and cultural elites, while the latter find it difficult to openly criticize the prevailing religiosity which often seems to suffuse the culture. This is probably a good sign that the system is actually working.

23. See J. Israel, "Spinoza, Locke, and the Enlightenment Battle for Toleration," supra n. 17. There is, as far as I know, no evidence that Spinoza's more radical ideas were known in eighteenth-century America.

CHAPTER 4

AN APPEAL TO HEAVEN: THE LANGUAGE OF RIGHTS ON THE EVE OF AMERICAN INDEPENDENCE

T.H. Breen

The scene is numbingly familiar. In urban centers one encounters ordinary men and women pouring into the streets. They demand greater freedom, an end to political oppression, a recognition of human equality, and the establishment of democratic procedures. Throughout restless and unstable societies, nongovernment organizations speak in the name of the people, seizing authority from autocratic regimes that seem to have lost legitimacy. Protesters champion the language of rights. Without working out the philosophic niceties of their own claims for due process and toleration, the downtrodden assert that simply because they are human beings they deserve the same liberty enjoyed by others who happen to control power. A demonstration of this sort could occur anywhere in the modern world. What contemporary Americans often forget is that the description also captures their own revolution, a violent regime change that began as a colonial rebellion against an empire of unquestioned military superiority.

Drawing attention to events that occurred more than two centuries ago may seem an enterprise of limited value for the peoples of the twenty-first century struggling against tyranny. After all, news reporters regularly explain that those who suffer oppression throughout the world view the United States with understandable skepticism. The dispossessed wonder whether Americans really are committed to universal human rights. For such men and women the run-up to revolution in 1776 may appear a narrative largely irrelevant to their immediate concerns. But it should not be so for modern citizens of the United States. As they accept the trappings of empire, they would do well to remember their own political origins, for once long ago, their ancestors appealed to the same bundle of human rights that inspires so many people today. If Americans return to the moment of independence, they may better comprehend the liberating possibilities of equality and freedom.

Why Americans need to be reminded of their own political heritage is a question of great significance. In an essay reconstructing the language of rights on the eve of national independence, it is impossible to do more than raise the issue. Even at the time of the American Revolution almost everyone believed that men and women possessed a loose, vaguely defined bundle of human rights. Disagreements surfaced when outspoken liberals such as Thomas Paine insisted that if human rights were indeed universal—a gift from God, not governments—then all people must enjoy them. The poor had rights. Black people had rights. Women had rights. As Paine and others quickly discovered, many Americans were not prepared to accept such an inclusive vision of rights, and although they mouthed rhetoric about equality, they rigorously excluded from the circle of rights those persons who threatened privilege and tradition.

As this piece argues, however, once a society takes on board the language of rights, it puts on the defensive those people who would seek to exclude others. Throughout the history of the United States, Americans have carried on a conversation amongst themselves not about rights in the abstract, but rather, about the boundaries for inclusion. Over the decades blacks, women, immigrants, and gays have powerfully proclaimed their human rights, the universalizing rights that first burst forth in 1776. As Americans face a troubled new millennium, they should appreciate that as representatives of a genuinely liberal polity, they cannot credibly deny others the opportunity to claim the same timeless rights for themselves.

I

On or about six o'clock on the morning of July 18, 1775 an event occurred which invites us to reconsider what we think we know about the role of ideas in the achievement of American independence.[1] Only a few months earlier the colonists had engaged British regulars at Lexington and Concord, and then later at Bunker Hill, and in the intoxicating wake of these successful encounters, colonial militia units rushed from communities throughout New England to the suburbs of Boston. George Washington had not yet arrived to take command of the American forces; a formal declaration of independence would not be issued for almost a year. During those heady days when protest turned to rebellion, rhetoric to violence, the main units of the Continental line serving under General Israel Putnam unfurled for the first time the standard that they intended to carry into battle. Several accounts of the dawn ceremony have survived. The fullest of these explained that the untested colonial troops assembled

> at Prospect-Hill, when the Declaration of the Continental Congress [of the necessity for taking up arms] was read; after which an animated and pathetic Address to the Army was made by the Rev. Mr. Leonard, Chaplain to General

Putnam's Regiment, and succeeded by a pertinent Prayer, when General Putnam gave the Signal, the whole Army shouted their loud Amen by three Cheers, immediately upon which a Cannon was fired from the Fort, and the Standard lately sent to General Putnam was exhibited [,] flourishing in the Air, bearing . . . this Motto, APPEAL TO HEAVEN.[2]

A loyalist spy confirmed the news, reporting to General Thomas Gage in Boston that "our people [have] got a famous New large Standard," and when it was raised "all of us huzzard [huzzahed] at once, then the Indeans [*sic*] gave the war hoop and to conclud[e], of[f] went the Cannon," a grand ceremony "that was worth you seeing."[3]

About the source of the army's chosen motto, there can be little doubt. The soldiers discovered the phrase in the pages of John Locke's *Second Treatise*, where "Appeal to Heaven" appears several times. Locke prudently associated these words with the story of Jephthah in the eleventh chapter of *Judges*, but references to the Old Testament hardly disguised the radical implications of Locke's conclusions. He counseled that in extreme cases when a ruler had forfeited the trust of the people and that people seeking justice found that the very judges constituted to hear their cause had compromised their independence by siding with the tyrant, the people had the right to take their political grievance to the Lord, in other words, to appeal to heaven. In the penultimate paragraph of the final chapter of the *Second Treatise* entitled the "Dissolution of Government," Locke explains,

> For in Cases where the Prince hath a Trust reposed in him, and is dispensed from the common ordinary Rules of the Law; there, if any Men find themselves aggrieved, and think the Prince act contrary to, or beyond that Trust, who [then, are] so proper to *Judge*[,] as the Body of the *People*, (who, at first, lodg'd that Trust in him) how far they meant it should extend? But if the Prince, or whoever they be in the Administration, decline that way of Determination, the Appeal then lies no where but to Heaven.[4]

So much of what Locke has to say about rights in the *Second Treatise* pertains to individuals, but here, we note, he specifically described the appeal as an act of desperation by a community, which after much suffering and deliberation, feels itself betrayed.

The Continental soldiers who justified their own political resistance through an "Appeal to Heaven" did not have rummage through musty university libraries to read Locke's words. Nor, did they have to rely on ministers such as the Reverend Leonard to tell them what the seventeenth-century theorist had written. A popular edition of the *Second Treatise* had just been issued by a Boston publisher who in a newspaper advertisement advised Americans,

> Perhaps there never was a Time since the Discovery of this new World, when People of all Ranks every where show'd so eager a Spirit of Inquiry into the Nature of their Rights and Privileges, as at this Day. This at all Times is a laudable

Spirit, and ought to be encouraged—It has therefore been judged very seasonable
and proper to put it in the Power of every free Man on this Continent to furnish
himself at so easy a Rate with the noble Essay just now re-published.

The advertisement guaranteed that "This Essay alone, well studied and
attended to, will give to every intelligent Reader a better View of the Rights
of Men and of Englishmen." Women as well as men were invited to study
the text. The American printer also noted that he completely dropped
Locke's *First Treatise*, a market decision that not only lowered the price of
the new edition, but also saved "all Lovers of Liberty" the nuisance of having to slog their way through a "prolix Confutation of Filmer and his
Disciples, few of which are yet to be found in this Country."[5]

The phrase "Lockean Moment" as used in the original title of this chapter is intended to call attention to a brilliantly inventive period—roughly the
years from 1768 to 1776—during which ordinary men and women in
America explored for the first time the full radical meaning of a language of
rights. To be sure, much of the public discussion owed a lot to the writings
of Locke himself. But, in a larger sense, Locke became a symbol for a wider,
often wonderfully undisciplined exploration of natural and human rights. It
was a moment during which a loose bundle of ideas about the contractual
foundations of civil authority suddenly and dramatically informed genuine
political resistance, a time when often inchoate beliefs and assumptions about
the state energized events, merging theory and practice, philosophy and politics, so that participants not only understood their own sacrifices within a
powerful intellectual framework, but also translated the language of rights
into an utopian vision for a better, freer, more egalitarian social order.[6]

My purpose in constructing a "Lockean Moment" specially around the
display of a Continental battle flag is two-fold. First, in most intellectual histories of the American Revolution it is precisely the people who actually
risked their lives in the cause of national independence who most often go
missing. We learn that men and boys entered the army for a host of highly
personal reasons: bad marriages, dreams of adventure, escape from overbearing fathers, and desire for extra money. Seldom do we interpret their
actions in relation to familiar narratives of political thought.[7] Apparently, that
kind of intellectual explanation is reserved for the privileged leaders, the
lawyers and planters who attended the congresses or served in the legislative
assemblies, and no one seems much inclined to inquire whether these more
prominent figures also had bad marriages, overbearing fathers, or dull lives
in isolated farm communities. Be that as it may, we should not forget that
without the support of the kind of men who celebrated a flag proclaiming an
"Appeal to Heaven," there would have been no independence. The
Revolution was, after all, the second bloodiest conflict in American history
in terms of the number of casualties in proportion to the total population,
and it would be a mistake either to take the mobilization of ordinary
colonists for granted, or worse, to condescend to these people by suggesting
that they were innocent of the ideas that shaped the new political culture.

The second reason why I have chosen to highlight a "Lockean Moment" has to do with the cultural construction of political memory, or as an anthropologist might say, with the political stories that modern Americans tell themselves about how they got to be they way they are. In other nations revolutionary history is a source of contestation, and various parties look to certain moments to legitimate their own current policies and aspirations. One only has to survey the literature of the French Revolution to appreciate my point, for in that country, one can provoke genuine anger by suggestions that the Terror as opposed to the summoning of Estates General represented the true spirit of revolution. For a very long time, the English Civil Wars could generate interpretive hostilities of this sort. But in the United States, the Constitution—its writing and its ratification—has come to dominate our memory of our own revolutionary past. We are instructed to see the years before 1776 as a necessary preparation for the creation of a republican structure of government that has managed to survive for more than two centuries. This perspective almost always discounts a radical language of rights and allows people made nervous by appeals to heaven, to declare the Constitution of 1787 to have been our real revolutionary achievement.[8] Unabashedly, I want to challenge such ideological claims and thereby, to remind a somewhat dispirited public that a historical language of rights— whatever its putative faults—may be the best thing on offer to revive a meaningful capacity for solidarity and resistance against global forces that have little use for talk of human rights.[9]

I do not mean to suggest that interpreters of the American Revolution have ignored the language of rights. Far from it. They have generally admitted that ordinary people on the eve of independence chattered incessantly about their rights, but then, in a series of quite extraordinary interpretive moves, the commentators have explained that the colonists got it all wrong. The Americans, we learn, did not understand the intellectual history of rights theory. They confused basic concepts, reducing their passionate claims about human and natural rights to a kind of nonsense that may have made sense in the taverns or meeting houses but hardly counted as a respectable body of ideas.[10]

British contemporaries were perhaps the first to point out how aggressively Americans mouthed shallow notions about rights. Dean Josiah Tucker, for example, found it almost impossible to take the colonists seriously in this regard, and in his *The Respective Pleas and Arguments of the Mother Country and the Colonies Distinctly Set Forth*, published in 1775, he weighed the solid "Facts and Precedents" advanced by the leaders of parliament against the Americans' airy rhetoric about "what they call immutable Truths—the abstract Reasonings, and eternal Fitness of Things—and in short to such Right of Human Nature which they suppose to be unalienable and indefeasible." Even at this early date, Tucker labeled the colonists "the Disciples of Mr. Locke," a description that suggested that they—like the Irish earlier in the century—had succumbed to a kind of ideological madness. To this characterization, Tucker's spirited adversary Richard Price responded, "Glorious

title!" But Tucker was not persuaded. In 1782 he was still grumbling about
the fuzzy-thinking Americans, and in a pamphlet entitled *Cui Bono?*, he
imagined what would happen should the provincial rebels actually emerge
from the war victorious. "When that happy Day should come," the Dean
observed with bitter sarcasm, "all Grievances, and all Complaints would
cease for ever. The People of *America* were to be blessed with a *Lockean*
Government, the only just one, the only free one upon Earth." Other
English writers took up the chorus, announcing as early as 1775—a year
before the Declaration of Independence—that the colonists had foolishly
appointed Locke "their political Apostle" and "their professed Director."
John Roebuck, another London pamphleteer, added more soberly in 1776
that the "present revolt has arisen solely from speculative notions" about
natural rights.[11]

One can appreciate why people like Tucker might have greeted rights talk
with such hostility. Less clear is why modern historians in the United States
have so frequently echoed the charges of these early critics. Their perspective, of course, is not the breakup of the first British Empire. Rather, I sus-
pect, their lack of enthusiasm for an earlier language of rights stems from a
conviction that arguments based on natural or human rights have little relevance to the most pressing social and political problems of our own times
and therefore, that it is not particularly useful or productive to return to an
earlier moment when so many ordinary Americans wove rights claims into
the rhetoric of personal and political liberation. Be that as it may, the bill of
particulars lodged against the language of rights is daunting. It was fundamentally incoherent, its detractors insist, promiscuously mixing charter
rights, legal rights, human rights, natural rights, and constitutional rights
into a conceptual muddle that strikes some intellectual historians as slightly
embarrassing, as if we had inadvertently allowed a bunch of secondary school
drop-outs to participate in a proper academic seminar on political theory.[12]
As one scholar recently observed of the colonists' use of Locke, "The cita-
tions are plentiful, but the knowledge they reflect . . . is at times superficial.
Locke is cited often with precision on points of political theory, but at other
times, he is referred to in the most offhand way, as if he could be relied on
to support anything the writers happened to be arguing."[13]

Historians of the American Revolution quickly provided an alternative
explanation for the role of ideas in bringing about national independence. In
what has become the new orthodoxy—indeed, an interpretation that has
reigned almost unchallenged for more than four decades—Locke plays
almost no constructive role, and we are told that however passionately the
colonists may have complained of violations of their rights, the driving ideology of revolution was an ancient discourse that appears in the current literature under various names such as "civic humanism" or "classical
republicanism." Whatever its label, this bundle of political ideas encouraged
late eighteenth-century colonists to imagine history as an incessant controversy between virtue and power, between liberty and corruption.[14] This conspiratorial view of human affairs, we are told, provided Americans with

a checklist against which they could measure the rise of tyranny and the decay of freedom. A standing army, state censorship of the press, a legion of grasping placemen perverting the legislative process, a rising national debt, the creation of new Anglican bishops—these were signs that all was not well in the commonwealth, and the only hope to preserve liberty against so many furious enemies was for independent, landed citizens to stand vigilant against the forces of ministerial corruption. It remains to be demonstrated whether the republican discourse—a kind of active, muscular citizenship—resonated positively among the ordinary people who actually took up arms against king and parliament, but we might observe in passing that this was a dark, negative, slightly paranoid ideology, one designed to inflame the fears and suspicions of marginal men and women without providing them with an ennobling invitation to remake the world within a framework of rights and equality.

The ascendance of the so-called republican discourse spawned a learned search for the origins of certain key political concepts. Even historians who professed themselves critical of an old fashioned genealogy of ideas industriously traced the development of a long conversation in Early Modern Europe and America about virtue and corruption, about independent citizenship and an ancient constitution. It was an enterprise in which texts talked to texts in an ever more abstract language. In his magisterial study *The Machiavellian Moment*, John Pocock provides an example of this type of approach to the history of political thought. He depicts the formative ideas of the revolutionary period as possessing "all the characteristics of neo-Harringtonian civic humanism.... The Whig canon and the neo-Harringtonians, Milton, Harrington, and Sidney, Trenchard, Gordon and Bolingbroke, together with the Greek, Roman, and Renaissance masters of the tradition as far as Montesquieu, formed the authoritative literature of this culture."[15] Another person surveying this field observes that the "reigning ideas of the American Revolution" were neither simple nor Lockean, "but complex and atavistic growing out of the rich English intellectual traditions of the Dissenters, radical Whigs, Classical Republicans, Commonwealthmen, Country party, or more simply, the Opposition."[16] With understandable impatience, this writer terms the scurrying after ever more obscure references to be an "ornate intellectual tradition." It is a fair point. However much one may admire this kind of history, one must admit that it takes us a long way from experience of the Continental soldiers who in 1775 actually made an impassioned appeal to heaven.

Something here has obviously gone amiss. The historical literature dissects an ideology that allegedly sparked revolution, and yet, from the perspective of what James Otis, a leader of protest in Massachusetts, called "vulgar" commonsense, it is hard to comprehend why young colonists would have found a bundle of conspiratorial ideas about virtue and power, about the supposed dangers of commercial society, sufficiently compelling to risk their lives against the strongest military force the world had ever seen.[17] Without getting bogged down in methodological considerations, I would submit that historians have made three assumptions about the role of ideas during the run-up to independence that have greatly impeded their appreciation of the language of rights in mobilizing the American people.

First, the analysis of political ideas has tended to rely on pamphlets, in other words, on formal, often well-documented presentations written for the most part by colonial lawyers and planters interested in impressing other lawyers and planters. There were exceptions, of course, and publications such as John Dickinson's *Farmer's Letters* reached a sizable audience. It did so, however, largely through the newspapers, which serialized Dickinson's writings, and not exclusively as a pamphlet. As Silas Deane, a Connecticut leader, observed in 1776, ordinary Americans were generally literate, and even "the very poorest" furnished themselves "with Gazettes...which they read, observe upon and debate...with a Circle of their Neighbors."[18] The point is not that the pamphlets should be scrapped in favor of newspapers. It is rather that the newspapers are closer to the ground as it were, and since it is in their pages that we encounter an exuberant celebration of rights, we should award them the same evidentiary standing as we do to the more cultivated productions.

The second problem relates to periodization. If we begin our discussion of the role of ideas with an assumption that the American Revolution included almost everything that occurred between 1760 and 1800, in other words, all the changes, crises, and contingencies that confronted the American people between the coronation of George III and the election of Thomas Jefferson, we are bound to lose sight of the specific contexts of specific events. I am not counseling a kind of micro-study that would treat each moment as somehow unique. It would, however, appear counter-intuitive to suppose that certain political ideas had the same purchase at one time as at another. Put a different way, the mobilization of a people for colonial rebellion in 1776 was not at all the same thing as convincing them to ratify a constitution in 1787; persuading men to confront the armies of Great Britain was a quite different enterprise from the need to legitimate political parties, which dominated public life during the 1790s. As one commentator recently observed,

> Shortly after the American Colonies won their War of Independence, their sovereignty was recognized by Britain, allowing them to get on with the internal talk of forging a cohesive nation. The new country quickly shifted its attention from self-justification to self-definition.[19]

This is a good point. If we insist that one political ideology fits all purposes, we shall find ourselves explaining too little by trying to explain too much.

A third problem—one obviously related to the other two—is that historians of eighteenth-century political thought in the United States have seldom given careful consideration to exactly what the ideas and beliefs that they analyze were supposed to achieve. Of course, it is perfectly reasonable to discuss concepts of liberty and justice, rights and authority from a timeless perspective that concentrates attention on the great theorists of the Western political tradition. My point is that at certain historical moments abstract principles about political life merged dramatically, even explosively, with real politics. When this occurred, they generated action, justified sacrifice, and invited ordinary people to consider with a sudden and exciting intensity exactly what sort of society they wanted for themselves as well as for their children's children. It is in these contexts—the period from 1768 to 1776, for example—that we discover a capacity among strangers for mobilization in a common cause and the creation of genuine political solidarities.[20]

These three general reflections return us to the "Lockean Moment." During the decade before the Continental Congress formally voted to declare independence, the language of rights reverberated through the press. It informed sermons and riveted the attention of town meetings. It is certainly true that when Parliament first began to tax the colonists in ways that they deemed unacceptable, Americans spoke of their rights in an unfocused way that seemed to confuse charter and statute rights with the more abstract claims of natural and human rights. That was to be expected. The initial instinct of the colonists was to demand the same rights that they assumed that they shared with all Englishmen, and many of these rights were historical in character. As the controversy intensified, however, it became increasingly clear that in the game of matching precedent with precedent, statute with statute, the colonists would lose. Quite literally, the context of the debate forced them to leap out of history, in other words, to stress God-given rights that were not dependent on Common Law, Magna Carta, or the Glorious Revolution. The Irish had made exactly the same interpretive move for similar reasons earlier in the century. In the case of the American colonists, a people without a shared sense of their own historical identity separate from that of the British Empire relied ever more tenaciously on timeless rights which owed little to legal and constitutional traditions.[21]

The eighteenth-century language of rights in revolutionary America consisted of several separate, though tightly related strands. To treat them independently is a somewhat artificial exercise, since in the popular political rhetoric of the day, they derived persuasive force from the assumption that one piece fit seamlessly into the others. Indeed, it is modern critics who have insisted on taking this or that concept out of context so that the colonists appear to have advocated a kind of hyper-individualism or a selfish sense of property which, in point of fact, they never did.[22]

The militiamen who flocked to Boston during the late spring of 1775 believed in the *universality* of human rights. In that sense they were children of the Enlightenment. Moreover, even though all rights derived ultimately from God, they were not a source of divine mystery. Any reasonable person could with only a minimum of reflection define his or her rights. Religious belief provided a firm philosophic foundation for the language of rights, and however difficult this may be to accept in a more secular society, the divinely-sanctioned character of rights meant that political morality was not simply a question of community sentiment. As Alexander Hamilton announced in 1775,

> the sacred rights of mankind are not to be rummaged for among old parchment
> or musty records. They are written, as with a sunbeam, in the whole *volume* of
> human nature, by the hand of divinity itself, and can never be erased or
> obscured by moral power.[23]

It is significant—if only because so many modern commentators choose to
ignore it—that rights talk in this context implied responsibility, mutual obligation, a respect for the rights of others. "Philanthrop" reminded the readers of the *Boston evening-post* in 1771 that when a man enters into a political
society, "he no longer considers himself an *individual*, absolutely unaccountable and uncontroulable, but as one of a community, every member of
which is bound to consult and promote the *general good*."[24]

Within this popular Lockean discourse, colonists insisted that their rights
were antecedent to the institution of government, and while positive laws
may have confirmed the existence of these rights, their authority did not
depend on the actions of judges or elected representatives. On this point
John Adams could hardly contain himself. "I say RIGHTS," he exclaimed,
"undoubtedly, antecedent to all earthly government,—*Rights*, that cannot
be repealed or restrained by human laws—*Rights*, derived from the great
Legislator of the universe."[25] The explicit point of such rhetoric was reminding colonists that any form of political authority—be it monarchy or republic—required the voluntary consent of persons who were by nature free and
equal and who surrendered as few of their antecedent natural rights as possible by submitting to the rule of law or government. As "Valerius Poplicola"
explained, "Every man was born naturally free, nothing can make a man
subject to any commonwealth, but his actually entering into it by positive
engagement, and express promise & compact."[26]

And finally, the colonists who decided to stand against the empire
believed that rights presupposed human equality, and even though many
Americans blatantly discriminated against women and blacks—just to cite
two obvious groups—they employed the word equality so unremittingly
that, as we shall see, some slaves interpreted this rhetoric as an invitation to
liberation. In 1773 the Reverend John Allen, a radical itinerant Baptist,
boldly instructed "the Right-Honourable the Earl of Dartmouth" about
American notions of natural equality. Allen wanted to know:

> As a fly, or a worm, by the law of nature has as great a right to liberty, and free-
> dom (according to their little sphere of life), as the most potent monarch on
> earth: and as there can be no other difference between your Lordship, and
> myself, but what is political, I therefore without any further apology, take leave
> to ask your Lordship, whether any one that fears GOD, loves his neighbour as
> himself...will oppress fellow-creatures?[27]

The nobleman did not respond. But other Americans found nothing objec-
tionable about Allen's assumptions. The Reverend John Tucker of Newbury,
Massachusetts, explained in an "Election Sermon" delivered in 1771 that "All
men are naturally in a state of freedom, and have an equal claim to liberty.

No one, by nature, not by any special grant from the great Lord of all, has any authority of another."[28]

As I noted earlier in this lecture, some historians who have discounted the centrality of the language of rights in the coming of the American Revolution draw attention to its muddled, vague, even incoherent character. And they are correct to do so. I would like to argue, however, that in this instance vagueness was a source of strength. Colonists could proclaim their God-given, universal, egalitarian rights without having to be too specific about which rights they had in mind. William Patten understood the problem. In a speech celebrating the repeal of the Stamp Act in 1766, he noted,

> As freedom and liberty are so comprehensive terms, it may be necessary to specify in what sense they are used in this discourse. When I say that every man is by nature free, I intend, that he has a natural right to life, to safety, to judge, determine and act for himself, to such things as are common, to enjoy what by the blessing of providence on his own industry, he has acquired.[29]

What these angry colonists discovered was that rights talk was fundamentally an emotive language which over time would prove itself intellectually unstable since there were always some people in society who wanted to claim natural or human rights that others insisted were utterly out of bounds. A tension between inclusion and exclusion—between those who enjoy certain rights and those who demand them on the basis of their undoubted humanity—is built into rights negotiations, and neither the failure to make good on universal equality nor the annoying propensity of rights talk to escalate into unacceptable extremes should deflect our attention from a "Lockean Moment" when the articulation of rights empowered ordinary people to reimagine political society.

And as men and women began to converse seriously about their rights, the key concepts of the discourse acquired a commonsensical quality, so that a phrase like an "Appeal to Heaven" triggered a bundle of popular assumptions about rights. This condition must have been what Jefferson had in mind when he responded to charges that his draft of the Declaration of Independence contained no original ideas. Richard Henry Lee had observed that Jefferson had relied entirely on "Locke's treatise on Government." Jefferson protested that he had not in fact reviewed Locke during the summer of 1776, but even so, he confessed that the language of rights was so commonplace, so much a part of the public discourse, that

> I did not consider it as any part of my charge to invent new ideas altogether.... My aim was simply to place before mankind the common sense of the subject....It was intended to be an expression of the American mind...the harmonizing sentiments of the day, whether expressed in conversation, in letters, printed essays, of the elementary books of public right."[30]

How exactly the language of rights became an "expression of the American mind" has bedeviled historians in the United States for a very long

time. Many take the long view—one championed among others by Alexis de
Tocqueville—and declare a boisterous commitment to individual rights to
have been a product of the earliest settler societies. Be that as it may, I shall
take a narrower perspective on the problem and draw attention to a curious
meeting that was held in Boston in November 1772. A shift in parliamentary
policy toward the colonies alarmed the town's elected officials, and they
gathered to address the latest imperial crisis. Their deliberations, however,
soon took an unexpected turn. Samuel Adams, a leader of the popular resistance in Boston, moved

> that a committee of correspondence be appointed, to consist of twenty-one
> persons, to state the rights of the Colonists and of this Province in particular,
> as men and Christians and as subjects; and to communicate and publish the
> same to the several towns and to the world as the sense of this town.

A local government assembly was suddenly transformed into a kind of
"teach-in," and with several hundred people in attendance, the group
explored the character of their rights long into the night.[31] When the Royal
Governor of Massachusetts heard what was happening, he grumbled that the
Boston committee consisted of a gang of "deacons," "atheists," "black-
hearted fellows, whom one would not choose to meet in the dark."[32]
Perhaps they appeared so to Thomas Hutchinson, but we should remember
that many of these "black-hearted fellows" would soon be serving in the
Continental Army.

The committee's final report was a mishmash of "vulgar" Lockean theory.
It contained precisely the type of confused rhetoric that drives modern historians of political thought to distraction. Adams and his friends first considered the "natural rights of the colonists as men." Most of their ideas came
directly from the *Second Treatise*. They then moved onto an examination of
"the rights of the colonists as Christians" and "the rights of the colonists as
subjects." At one point, the Town Meeting concluded—again in classic
Lockean terms:

> it is the greatest absurdity to suppose it in the power of one or any number of
> men at the entering into society, to renounce their essential natural rights, or
> the means of preserving those rights when the great end of civil government
> from the nature of its institution is for the support, protection and defense of
> those very rights.[33]

If the committee's work had gathered dust in a volume of the official
records of the Boston Town meeting, it would be of some, though limited
value to a social historian of political ideas. It did not do so. Samuel Adams
insisted on dispatching copies to the scores of inland villages that were just
then trying to make sense of impending crisis. The thoughts of Boston
"black-hearted" men circulated widely, provoked new discussion, and challenged ordinary people to decide just how far they were prepared to go in
the defense of their rights.[34] A town meeting in Marblehead voted that "the

pamphlet containing the state of rights, &c., be lodged in the town clerk's office, and read annually at the opening of every March meeting for the election of town officers, until the public grievances be redressed."[35] On February 10, 1773 the voters of Mendon passed as series of resolutions, the first of which was "that all Men have naturally, an equal Right, to Life, Liberty and Property." They moved logically step by step through Locke's theory of government and resistance, concluding that any effort by parliament to curtail the natural rights of the colonists was "not reconcilable to the most obvious Principles of Reason; as it subjects us to a State of Vassalage, and denies us those essential Natural Rights, which being a Gift of GOD-ALMIGHTY, it is not in the Power of Man to alienate."[36] The Mendon voters did not adopt the phrase "an appeal to heaven," but they surely understood its growing relevance to their political future. And in the tiny farming community of Harvard, located some miles to the northeast of Worcester, the townsmen decided on March 1, 1773,

> That the great end of government and of men's entering into Civil Society is the preservation and security of their just Rights, Liberties and Properties against unjust force and violence, and when this is disregarded by those who govern, they Counteract the very Design for which Government was Instituted, and the governed have a right to judge of what is for the benefit of the Society as well as the Governour.[37]

These records were in no way unusual. Like the Putney Debates during the English Civil Wars, these community gatherings brought political theory into creative contact with real politics. These people had a lot to lose. No doubt, had the language of republican virtue and power served their immediate purposes, the townsmen of Marblehead, Mendon, and Harvard would have cited different sources, make less of rights talk. But they did not do so.

Nor significantly did the enslaved black people of New England choose to do so. These men and women provide a remarkable opportunity to understand better the political thinking of the mainstream white society on the eve of national independence. This claim may sound a little perverse. After all, historians of eighteenth-century slavery generally concentrate on reconstructing the experiences of the blacks themselves—as slaves—and if these people figure at all in discussions of political ideology, they generally serve as reminders that however much white colonists objected to their own loss of liberty, they did not do much to restore the freedom of the blacks.[38] But for my purposes today, I am not much interested in how white colonists perceived the blacks. I take the opposite perspective. The enslaved blacks of New England realized that after 1770 the political climate was fast changing and that they might successfully appeal to white colonial officials for freedom. It was a delicate situation. The blacks had to state their case in a form that would have the greatest chance of persuading their masters to liberate them from bondage.[39]

And, at the crucial moment, they knew exactly what arguments were likely to be most effective. They had listened in the taverns; they had heard whites

talking in the streets. When Caesar Sarter, a black man living in Essex County
north of Boston, drew up a petition for the freedom of the slaves in August
1774, he began quite appropriately with the language of rights: "this is a
time of great anxiety and distress among you [the whites], on account of the
infringement, not only of your Charter rights, but of the natural rights and
privileges of freeborn men." After reminding the white readers of the *Essex
Journal and Merrimack Packet* of the horrors of enslavement, he inquired,
"why will you not pity and relieve the poor distressed, enslaved Africans?—
Who, though they are entitled to the same natural rights of mankind, that
you are, are, nevertheless groaning in bondage."[40] In June 1773 a group of
blacks appealed to Governor Hutchinson—not someone likely to lend a sympathetic ear—"Your Petitioners apprehend they have in comon with other
men a naturel right to be free and without molestation to injoy such property as they may acquire by their industry, or by any other means not detrimental to their fellow men."[41] Like the ordinary farmers of Mendon and
Harvard, these blacks might have adopted a different language; they might
have begged the independent and virtuous republicans of the commonwealth to show compassion. But instead they celebrated the Lockean
Moment. Listen to a marvelous petition sent on May 25, 1774 to General
Thomas Gage, then head of the British forces in Boston, by "a Grate
Number of Blackes of the Province . . . held in a state of Slavery within a free
and Christian Country." "Your Petitioners," they wrote,

> apprehind we have in common with all other men a naturel right to our free-
> doms without being deprived of them by our fellow men [,] as we are a free-
> born Pepel and have never forfeited this Blessing by aney compact or
> agreement whatever. . . . We[,] therefore[,] Bage your Excellency and Honours
> will . . . cause an act of the legislative to be pessed that we may obtain our
> Natural right[,] our freedoms[,] and our children to be set at lebety."[42]

As Caesar Sarter and other obscure people would soon discover, a
Lockean Moment of such explosive intensity could not be sustained. During
the 1780s leaders of the new republic brought rights talk under tight control. They feared the incendiary possibilities of the language of rights, for if
every person could appeal to concepts of natural and human rights whenever
he or she so desired, the civil order might dissolve into anarchy. As John
Adams recognized as early as 1776, rights were disruptive. He sensed, for
example, that the slightest relaxation of franchise rights would open the door
to more radical demands for political participation. "New claims will arise,"
Adams predicted, "women will demand a vote; lads from twelve to twenty-one
will think their claims not closely attended to; and every man who has not a
farthing will demand an equal vote with any other, in all the acts of the
state."[43] And so rights were domesticated, written down, ordered, enumerated, and constrained by procedure. All of this was perfectly predictable. For
a post-revolutionary society the major challenges were drafting a constitution,
strengthening the bonds of national identity, and achieving fiscal stability.[44]

In this quite different ideological context even people like Jefferson began to express second thoughts about rights. To be sure, he still claimed to believe that all men were created equal and were endowed by their creator with certain unalienable rights, but after 1780 he carefully explained that since the blacks were not fully human, they could not possibly share the same natural rights as did the white yeomen farmers who figured so prominently in his *Notes on the State of Virginia*.[45]

In these matters we should not take Jefferson as our guide. Over the course of the history of the United States people like him have discovered that rights cannot easily be curtailed. They rattle around in the public memory, inviting the most unlikely collection of persons that one would never want to meet in the dark to step forward to claim their rights, to insist on their equality, and to demand full inclusion in civil society. The language of rights spoke powerfully to Abraham Lincoln and Martin Luther King, Junior. And at the commencement of a new millennium, we can anticipate other Lockean Moments. As we do so, we should keep alive a political memory of a time when ordinary men and women spoke in precisely these terms, and however tempting it may be for some in our society to affirm the view that the concepts expressed in the Declaration of Independence are vastly overrated, we must not do so. The stakes are too high.[46] We might remember an incident that Josiah Quincy, a young Massachusetts lawyer, reported during a trip to London on the eve of revolution. On January 20, 1775 he witnessed a debate over American affairs in the House of Lords. Lord Gower rose to rebut a speech just given by a peer who favored reconciliation with the colonists. According to Quincy, "His Lordship said I am for enforcing these [coercive] measures (and with a great sneer and contempt); [added] let the Americans sit talking about their natural and divine rights, their rights as men and citizens; their rights from God and nature."[47] And that is just what they did. If nothing else, Lord Gower and his allies—then and now—invite us to celebrate our enduring capacity as a people to preserve our rights through "appeals to heaven."

NOTES

An earlier version of this essay was published by Oxford University Press as *The Lockean Moment: The Language of Rights of on the Eve of the American Revolution* (Oxford, 2002) and is republished here with permission.

1. I want to thank Michael Guenther and Christopher Beneke of Northwestern University for providing research support.
2. *New-England Chronicle or Essex Gazette*, July 13–21. George Washington later commissioned a very similar flag to fly over two shore batteries and six armed vessels operating near Boston. See Edward W. Richardson, *Standards and Colors of the American Revolution* (Philadelphia: University of Pennsylvania Press, 1982), 59; and Alfred Morton Cutler, *The Continental 'Great Union' Flag* (Somerville, MA: City of Somerville, 1929), 27–28.
3. Cited in Henry Belcher, *The First American Civil War: First Period, 1775–1778* (London: MacMillan and Company, 1911), I, 208.

4. John Locke, *Two Treatises of Government*, ed. By Peter Laslett (Cambridge: Cambridge University Press, 1960), 427, also 282, 379–80, 386, and 428. Locke's radical phrase echoed curiously through the political debate in Massachusetts over parliamentary sovereignty. On July 31, 1770 the House of Representatives lectured the colony's Lieutenant Governor Thomas Hutchinson on the writings of "the great Mr. Locke." Specifically, the representatives quoted from the *Second Treatise*, arguing that in certain constitutional disputes between the people and their rulers, "The people have no other remedy...but to appeal to heaven." The representatives seemed content in this case merely to suggest where the conflict over taxation might lead, for at the end of the Locke quotation, they added, "We would, however, by no means be understood to suggest, that this people have occasion at present to proceed to such extremity." Hutchinson was not amused. He responded to the House, "Your quotation from Mr. Locke, detached as it is from the rest of the treatise, cannot be applied to your case. I know of no attempt to enslave or destroy you, and as you, very prudently, would not be understood to suggest that this people have occasion to proceed to such extremity as to appeal to heaven, I am at a loss to conceive for what good purpose you adduce it." Thomas Hutchinson, *The History of the Colony and Province of Massachusetts-Bay* (Cambridge, MA: Harvard University Press, 1936), III, 388, 395.
5. *Boston Gazette*, March 1, 1773.
6. Some of these themes are explored in T.H. Breen, "Subjecthood and Citizenship: The Context of James Otis's Radical Critique of John Locke," *New England Quarterly* 71/3 (1998), 378–404.
7. Two notable exceptions to this generalization are Charles Royster, *A Revolutionary People at War: The Continental Army and American Character, 1775–1783* (New York: W.W. Norton, 1979); and John Shy, *A People Numerous and Armed: Reflections on the Military Struggle for American Independence* (New York: Oxford University Press, 1976).
8. See, for example, Robert H. Bork, *Slouching Towards Gomorrah: Modern Liberalism and American Decline* (New York: Regan Books, 1997), 56–67.
9. For a persuasive and passionate discussion of the centrality of rights in contemporary American society, see Charles L. Black, Jr., *A New Birth of Freedom: Human Rights, Named and Unnamed* (New Haven: Yale University Press, 1997); also Jeffrey N. Wasserstrom, Lynn Hunt, and Marilyn B. Young, eds., *Human Rights and Revolutions* (Lanham, MD: Rowman and Littlefield, 2000).
10. For example, Phillip Reid, *Constitutional History of the American Revolution: The Authority of Rights* (Madison: University of Wisconsin Press, 1986); Also see Jack P. Greene, *All Men Are Created Equal: Some Reflections on the Character of the American Revolution* (Oxford: Oxford University Press, 1976) and Stanley N. Katz, "The Strange and Unlikely History of Constitutional Equality," *Journal of American History* 75/3 (1989), 747–63. The most satisfactory discussion of the language of rights on the eve of independence is Daniel T. Rodgers, *Contested Truths: Keywords in American Politics Since Independence* (Cambridge, MA: Harvard University Press, 1987), ch. 2.
11. Josiah Tucker, *Respective Pleas* (London, 1775), 16, no. 39; Richard Price, *Observations on the Nature of Civil Liberty...* in Bernard Peach, ed., *Richard Price and the Ethical Foundations of the American Revolution* (Durham,

NC: Duke University Press, 1979), 113; Tucker, *A Cui Bono?* (London, 1782), 94; [Anonymous], *Americans Against Liberty: Or An Essay on the Nature and Principles of True Freedom*...(London, 1775), 37, 46; and John Roebuck, An Enquiry Whether the Guilt of the Present Civil War.... (London, 1776), 67. In his *Letters on the Present Disturbances in Great Britain and her American Provinces* (London, 1777), Allan Ramsay assured English readers that "That People of England will be able to see through all the sophistry of the American Pamphleteers, who, having no sense of their own, borrow some from Locke." (24) The heated dispute in England over Locke and the American Revolution is recounted in Isaac Kramnick, *Republicanism and Bourgeois Radicalism: Political Ideology in Late Eighteenth-Century England and America* (Ithaca, NY: Cornell University Press, 1990), 163–99.

12. Rodger, *Contested Truths*, 50–53; Stephen A. Conrad, "Putting Rights Talk in its Place: *The Summary View* Revisited," in Peter S. Onuf, ed., *Jeffersonian Legacies* (Charlottesville, VA: University of Virginia Press, 1993), 254–80.

13. Bernard Bailyn, *The Ideological Origins of the American Revolution: Enlarged Edition* (Cambridge, MA: Harvard University Press, 1992), 28.

14. A summary of this literature can be found in Milton M. Klein, Richard D. Brown, and John B. Hench, eds., *The Republican Synthesis Revisited: Essays in Honor of George Athan Billias* (Worcester, MA: American Antiquarian Society, 1992); Also see Bailyn, *Ideological Origins*; Gordon S. Wood, *The Creation of the American Republic, 1776–1787* (Chape Hill, NC: University of North Carolina Press, 1969); and Michael P. Zuckert, *Natural Rights and the New Republicanism* (Princeton: Princeton University Press, 1994). Daniel T. Rodgers provides a sensible assessment of the reception of republican ideas in late eighteenth-century America and their relation to recent historiography in "Republicanism: The Career of a Concept," *Journal of American History* 79/1 (1992), 11–38.

15. J.G.A. Pocock, *The Machiavellian Moment: Florentine Political thought and the Atlantic Republican Tradition* (Princeton: Princeton University Press, 1975), 506–07.

16. In her own *Liberalism and Republicanism in the Historical Imagination* (Cambridge, MA: Harvard University Press, 1992) 161, Joyce Appleby distances herself from an increasingly arid search for origins and influences in intellectual history. Also valuable in a continuing reassessment of the role of ideas in the coming of revolution is James T. Kloppenberg, *The Virtues of Liberalism* (New York: Oxford University Press, 1998), 21–37 Older, but useful studies that stressed the importance of Locke in American thinking are Morton White, *The Philosophy of the American Revolution* (New York: Oxford University Press, 1978) and Carl L. Becker, *The Declaration of Independence: A Study in the History of Political Ideas* (New York: Vintage Books, 1942).

17. Cited in Bailyn, *Ideological Origins*, 186.

18. Cited in Robert M. Wier, "The Role of the Newspaper Press in the Southern Colonies on the Eve of the Revolution: An Interpretation," in Bernard Bailyn and John B. Hench, eds., *The Press and the American Revolution* (Worcester, MA: American Antiquarian Society, 1980), 100. On the widespread literacy of the colonists, see Kenneth A. Lockridge, *Literacy in Colonial New England, An Enquiry into the Social Context of Literacy in the Early Modern West* (New York: WW. Norton, 1974).

19. Ruth R. Wisse, "The Jewish State: The Struggle for Israel's Soul," *Times Literary Supplement* March 2, 2001, 3. The most thoughtful discussion of this profound shift in ideological perspective, one closely associated with the development of a post-independence national identity, can be found in Winthrop D. Jordon, *White Over Black: American Attitudes Toward the Negro, 1550–1812* (New York: WW. Norton and Company, 1977), 315–582.

20. In "Narrative of Commercial Life: Consumption, Ideology, and Community on the Eve of the American Revolution," *William and Mary Quarterly*, 3rd ser., 50/3 (1993), 471–501, I have explored the dynamics of popular mobilization during this period. Also Rodgers, *Contested Truths*, 50–57.

21. T.H. Breen, "Ideology and Nationalism on the Eve of the American Revolution: Revisions Once More in Need of Revising," *Journal of American History* 84/1 (1997), 13–40.

22. An example of this kind of ahistorical analysis is C.B. Macpherson, *The Political Theory of Possessive Individualism: From Hobbes to Locke* (Oxford: The Clarendon Press, 1962).

23. *Papers of Alexander Hamilton*, ed. By Harold C. Syrett et al. (New York: Columbia University Press, 1961–87), I, 122.

24. January 14, 1771.

25. *Works of John Adams*, ed. Charles F. Adams (Boston: Little, Brown, and Company, 1850–1856), III, 449.

26. *Boston Gazette*, October 23, 1771.

27. John Allen, *An Oration Upon the Beauties of Liberty* (New London, 1773) reprinted in Ellis Sandoz, ed., *Political Sermons of the American Founding Era, 1730–1805* (Indianapolis: Liberty Press, 1991), 305; Also see *Massachusetts Spy*, August 18, 1774.

28. John Tucker, *A Sermon Preached at Cambridge.* . . . (Boston, 1771) reprinted in Charles S. Hyneman and Donald S. Lutz, eds., *American Political Writings During the Founding Era, 1760–1805* (Indianapolis: Liberty Press, 1983), I, 162.

29. William Patten, *A Discourse Delivered at Halifax in the County of Plymouth, July 24, 1766*. . . (Boston, 1766), 7. A valuable discussion of the specific context of the language of rights during this period can be found in Philip A. Hamburger, "Natural Rights, Natural Law, and American Constitutions," *Yale Law Review* 102/4 (1993), 907–60. Also Michael J. Lacey and Knud Haakonssen, eds., *A Culture of Rights: The Bill of Rights in Philosophy, Politics, and the Law—1791 and 1991* (Cambridge UK: Cambridge University Press, 1991).

30. Cited in Becker, *Declaration of Independence*, 25–26. Pauline Maier demonstrates just how much the Declaration reflected the political assumptions of a broader popular culture in *American Scripture: Making the Declaration of Independence* (New York: Alfred A. Knopf, 1997).

31. William V. Wells, *The Life and Public Services of Samuel Adams* (Boston: Little Brown, and Company, 1865), I, 496–97.

32. Cited in ibid., I, 497.

33. Ibid., I, 502–07.

34. Ibid., II, 4. See Richard D. Brown, *Revolutionary Politics in Massachusetts: The Boston Committee of Correspondence and the Towns, 1772–1772* (Cambridge, MA: Harvard University Press, 1970) and William M. Fowler, Jr., *Samuel Adams: Radical Puritan* (New York: Longman, 1997).

35. Wells, *Samuel Adams*, II, 4.

36. *Boston Gazette*, June 7, 1773.

37. Cited in Henry S. Nourse, *History of the Town of Harvard, Massachusetts, 1732–1893* (Harvard, MA: Warren Hapgood, 1894), 304–05.

38. Two of the better recent examples are Ira Berlin, *Many thousands Gone: The First Two Centuries of Slavery in North America* (Cambridge, MA: Harvard University Press, 1998); and Philip D. Morgan, *Slave Counterpoint: Black Culture in the Eighteenth-Century Chesapeake and Lowcountry* (Chapel Hill, NC: Omohundro Institute of Early American History and Culture, 1998).

39. I have discussed the radical political climate that encouraged colonial slaves to imagine liberation in "Making History: The Force of Public Opinion and the Last Years of Slavery in Revolutionary Massachusetts," in Ronald Hoffman, Mechal Sobel, and Fredrika J. Teute, eds., *Through a Glass Darkly: Reflections on Personal Identity in Early America* (Chapel Hill, NC: Omohundro Institute of Early American History and Culture, 1997), 67–95; and "A Monument for Barre: Memory in a Massachusetts Town," in William E. Leuchtenburg, ed., *American Places: Encounters with History: A Celebration of Sheldon Meyer* (New York: Oxford University Press, 2000), 29–40; Also Benjamin Quarles, *The Negro in the American Revolution* (New York: W.W. Norton, 1973); and Gary B. Nash, *Race and Revolution* (Madison, Wisc.: Madison House, 1990).

40. Cited in Roger Bruns, ed., *Am I Not a Man and a Brother: The Antislavery Crusade of Revolutionary America, 1688–1788* (New York: Chelsea House, 1977), 338, also 428, 452–53.

41. Cited in Sidney Kaplan, *The Black Presence in the Era of the American Revolution, 1770–1800* (Washington, DC: National Portrait Gallery, 1973), 13 [original spelling]; also see *Boston Gazette and Country Journal*, August 23, 1773.

42. Ibid. [original spelling.]

43. John Adams to James Sullivan, 26, 1776, *Works of John Adams*, IX, 375–76. For a discussion of the rights of women, see *Massachusetts Gazette and Boston Post-Boy*, May 30, 1774.

44. Rodgers, *Contested Truths*, 53–66; and Breen, "Ideology and Nationalism," 13–40. On force of nationalism after the Revolution, see David Waldstreicher, *In the Midst of Perpetual Fetes: The Making of American Nationalism, 1776–1820* (Chapel Hill, NC: University of North Carolina Press, 1997); Also see Michael Zuckerman, "A Different Thermidor; The Revolution Beyond the American Revolution," in James A. Henretta, Michael Kammen, and Stanley N. Katz, eds., *The Transformation of Early American History: Society, Authority, and Ideology* (New York: Alfred A. Knopf, 1991), 170–93.

45. The most impressive discussion of the relation between nationalism after independence and the appeal of scientific racism and of Jefferson's part in that development remains Jordan, *White Over Black*.

46. Black, *A New Birth of Freedom*; Ronald Dworkin, *Taking Rights Seriously* (Cambridge, MA: Harvard University Press, 1977); Robert Westbrook, "Comment," *Radical History Review* 71/1 (1998), 46–51; and Thomas Haskell, "The Curious Persistence of Rights Talk in the 'Age of Interpretation,'" *Journal of American History* 74/3 (1987), 984–1012.

47. "Journal of Josiah Quincy, Jun., During His Voyage and Residence in England From September 28th, 1774, to March 3rd, 1775," *Proceedings*, Massachusetts Historical Society, 50 (1917), 463.

CHAPTER 5

JEFFERSON AND JINNAH: HUMANIST IDEALS AND THE MYTHOLOGY OF NATION-BUILDING

Akbar S. Ahmed

INTRODUCTION

On September 11, 2001, the world of Thomas Jefferson, founding father of the United States, and that of Muhammad Ali Jinnah, the founding father of Pakistan, came dramatically and tragically face to face with each other. The events of that day would impact on both societies in ways that the respective founding fathers could not imagine. America launched a war against terrorism. The immediate target was the Taliban regime and Al-Qaeda with its leader Osama bin Laden in Afghanistan. Considering that the Taliban were nurtured and educated in neighboring Pakistani schools, and that the movement has been supported by Pakistan, the role of that country in America's war against terrorism became crucial.

Examining Jefferson and Jinnah becomes instructive when analysing our times. Indeed, more than instructive, an examination of their lives becomes inspirational, as it assists the rediscovery of their ideas relating to the kind of society and the sort of individual they envisaged. The rediscovery will allow us to understand what ignited the social and political movements that embodied the ideas of liberty, democracy, and equality.

There is a moment in the life of a people when an idea is translated into a movement that goes on to transform the people into a nation. Scholars in the West have used the phrase a "Lockean Moment." As one scholar so eloquently describes the moment:

> It was a moment during which a loose bundle of ideas about the contractual foundations of civil authority suddenly and dramatically informed genuine

political resistance, a time when often inchoate beliefs and assumptions about
the state energized events, merging theory and practice, philosophy and politics, so that participants not only understood their own sacrifices within a powerful intellectual framework, but also translated the language of rights into a
utopian vision for a better, freer, more egalitarian social order. (Breen 2001, 5)

Jefferson and Jinnah are both part of the Lockean moment in the history of
the respective nations they helped to create. They helped ignite what became
a movement for nationhood and in that early act of creation have come to
embody the nation itself.

THESIS

After September many urgent questions were thrown up in the media for
answers: Was Islam, the religion that the hijackers represented, incompatible
with democracy? Were there any democratic leaders in Islam? If there are,
what do they think of religious freedom, individual liberty, and the role of
education, values that have come to be identified as "humanist" deriving
from the Greeks? Indeed, are humanist ideals the exclusive preserve of
Western civilization? Can we discern similar ideals in Islam? Is there hope of
dialogue between these civilizations or is only a clash inevitable? I will
approach the answers to these questions by an examination of Thomas
Jefferson, but select one that allows me to compare him to Mr. Jinnah and
relate the two leaders, their history, and their culture.

They are two leaders working in two different cultural contexts on two
different continents and yet both echo ideas of liberty, religious freedom,
and the importance of education. Does their example suggest that these
ideas are universal, transcending culture, religion, and nation?

Given the fact that the two leaders come from such different cultural and
historical backgrounds, can a valid comparison be made? Does the belief of
both in the ideals of liberty, the inalienable rights to religious freedom, life,
and property suggest a meeting of minds transcending time and space?

In this chapter I argue that ideas of individual liberty, religious freedom,
and respect for universal education are not exclusive to Western civilization
but also to be found in Islamic civilization. I illustrate this through the comparison between Jefferson and Jinnah. I also illustrate why such ideas grew
in one society and were thwarted in another. My thesis then is that while the
individual leaders represent humanist ideals, the sociological dispensation of
their society determines the possibility of implementing or developing them.

An examination of Jefferson and Jinnah will raise several broad points
relating to the two cultures they represent: there is an implicit ethnocentric-
ity in equating ideas of democracy and liberty to Western thinkers starting
from the eighteenth century onward. Yet we know that Jinnah's notions of
justice, freedom, and democracy were consciously derived from Islamic history going back to the seventh century. We also note that the founding
fathers of America were just that—founding fathers. There were no mothers.

The Pakistan movement in comparison had its founding fathers, but also the towering personality of Fatimah Jinnah, who became the founding mother of Pakistan. Indeed in Pakistan she is known as *Madr-e-Millat*, or the "Mother of the Nation."

ANALYSIS

Of course Jefferson is not the only founding father of the United States. Perhaps a truer comparison with Jinnah would have been between him and George Washington. Both Washington and Jinnah became the official first heads of new nations. Both illustrated their respect for democracy: Washington by refusing to stand for a third term and Jinnah by giving up the powerful post of head of the Muslim League Party that created Pakistan. Both refused offers of kingship by enthusiastic supporters.

However, the intellectual and philosophic founding father of America was Jefferson. His contribution to the Declaration of Independence is unchallenged. His definition of political liberty, religious freedom, and the importance of education are woven into the self-definition of America. Similarly, Jinnah's first two speeches to the Constituent Assembly in August 1947 are crucial to understand his thinking. But while America treasured the ideas of Jefferson and they became part of cultural mythology, the ideas of Jinnah were sometimes lost to history and sometimes even dropped from it. His first speech to the Constituent Assembly in which he declared that Muslims would be free to go to their mosques and Hindus to their temples "disappeared" from the official records during certain periods of Pakistan's history. The rulers of Pakistan did not want to be reminded of Jinnah's tolerant vision of the state.

However, Jinnah was not unique. His intellectual lineage as a modern Muslim took him back to Sir Sayyid Ahmed Khan in the mid-nineteenth century. It was a time of great upheaval when Muslim society was looking for answers to the invasive presence of British modernity. Sir Sayyid's creation of Aligarh, which became a synthesis between Western and Islamic cultures, inspired generations of Muslims. Others like Allama Iqbal, educated at Cambridge and Heidelberg, also sought to raise the discussion of democracy and Muslim politics. Jinnah led the movement for Pakistan, the most significant Muslim political movement in the twentieth century.

After all, Jinnah eventually represented and lead the Muslims of South Asia who have played out their own history against the colonial power, the British. But it is a history that is delayed in time compared to the American War of Independence against the British. About a century and a half separate the two movements. Jinnah himself was born at the end of the nineteenth century; Jefferson died in the early half of the century. Given the similarity of ideas for society and nation, the question then is: How effectively were these ideas perpetuated and maintained in the nation?

For the United States the record is one of moving from strength to strength in the realization of a society dedicated to Jefferson's ideals. It was not always easy. Just decades after Jefferson died the Civil War tore the new nation apart and the healing process would take many generations. Americans were fighting in dramatic ways for civil liberties until the 1960s. Several prominent Americans lost their lives in the passion around ideas of American identity—from the two Kennedy brothers to Malcolm X and Martin Luther King.

September 11 again created dilemmas in society. Jefferson would have balked at the talk of secret evidence, wire-tapping, arrest without evidence, and the casual approach to the notion of *habeas corpus*. The Muslim community in America, feeling itself a target, would have invoked Jeffersonian ideals of liberty. There have been far too many complaints from the Muslim community. The actions of Osama bin Laden challenged the very heart of the Jeffersonian ideal.

Jinnah's Pakistan was also challenged by the actions of Osama bin Laden. The war against terrorism allowed President Pervez Musharraf—he had recently elevated himself to the Presidency from simply being the commander in chief of the Pakistan army—to declare a state of emergency. People were picked up to be locked away and civil liberties, in any case not very strongly developed, appeared to vanish. Politicians demanding elections were either banned or their leaders arrested, or harassed by Income Tax Officials. It was easy to label people as "terrorists" and do away with them. Clearing the way for military rule, Musharraf announced a questionable referendum and stayed on for another five years. Thus Pakistan entered the new millennium under a military dictator. Jinnah's dream of a democratic Pakistan was once again lost.

COMPARISONS

Sam Waterston playing the American presidential candidate in "Mindwalk" (1990) compares Thomas Jefferson to Isaac Newton. He places Jefferson at the core of the American political universe, like Newton is at the core of Western intellectual development explaining the very laws of modern science.

Thomas Jefferson is the quintessential founding father of the United States of America. Through what he said and did, he symbolizes the American ideal. An early biographer wrote of him: "If Jefferson was wrong, America is wrong."

An eminent Englishman who had worked in British India described Mohammed Ali Jinnah as the Pakistani equivalent of the Archbishop of Canterbury, the Prime Minister, the Speaker, and the King-Emperor all rolled into one.

There are interesting personal comparisons between Jefferson and Jinnah to be made apart from the striking similarity of ideas. Both were self-made men not belonging to any aristocratic lineage; both found it difficult to have comfortable personal relations with people around them; both were involved in relationships that raised eyebrows (Jefferson and his slave girl; Jinnah and his bride half his age); both were accused by their critics of inconsistency

(Jefferson for not being robust in defending Virginia from an invading British fleet with Benedict Arnold in command; Jinnah for abandoning his role as ambassador of Hindu–Muslim unity and becoming the champion of Pakistan); both were cerebral and not sensual men; both are cast in heroic terms by their followers as fathers of the nation and attract people from across the political spectrum; both men are remembered in national monuments which are on the visiting list of tourists (the Jefferson memorial in Washington, DC and Jinnah's mausoleum in Karachi).

Both Jefferson and Jinnah reflect complex psychological attitudes towards the aristocracy of their times which they challenge and yet wish to co-opt for their cause. Not quite from the upper reaches of the aristocracy in terms of lineage, they nonetheless live and behave as aristocrats. Both marry women who belong to the aristocracy. Yet both are strongly conscious of speaking on behalf of "the people." One of the great paradoxes of politics is the popularity of both Jefferson and Jinnah in the public mind in spite of their aloof, aristocratic and somewhat cerebral appearance.

Jefferson's last letter ever, written to the Mayor of Washington, on June 24, 1826 weaves the different strands of his thinking. His main concern is to create a society free of the tyranny of the ruling elite, those he calls "a favored few":

> May it [the Declaration of Independence] be to the world, what I believe it will be, (to some parts sooner, to others later, but finally to all,) the signal of arousing men to burst the chains under which monkish ignorance and superstition had persuaded them to bind themselves, and to assume the blessings and security of self-government. That form which we have substituted, restores the free right to the unbounded exercise of reason and freedom of opinion. All eyes are opened, or opening, to the rights of man. The general spread of the light of science has already laid open to every view the palpable truth, that the mass of mankind has not been born with saddles on their backs, nor a favored few booted and spurred, ready to ride them legitimately, by the grace of God. (Jefferson 1984, 1516–17)

Jinnah too warned against creating a nation that would allow the ruling class to perpetuate its hold over ordinary people:

> Here I should like to give a warning to the landlords and capitalists who have flourished at our expense by a system which is so vicious, which is so wicked and which makes them so selfish, that it is difficult to reason with them. The exploitation of the masses has gone into their blood. They have forgotten the lesson of Islam. . . . There are millions and millions of our people who hardly get one meal a day. Is this civilization? Is this the aim of Pakistan? [Cries of "No, No".] . . . If that is the idea of Pakistan, I would not have it. . . . The minorities are entitled to get a definite assurance or to ask, "Where do we stand in the Pakistan that you visualize?" (Merchant 1990, 10–11)

In 1946 he repeated the same theme in Calcutta:

> I am an old man. God has given me enough to live comfortably at this age. Why would I turn my blood into water, run about and take so much trouble? Not

for the capitalists surely, but for you, the poor people. . . . I feel it and, in
Pakistan, we will do all in our power to see that everybody can get a decent
living. (R. Ahmed 1993, 62)

Jinnah spoke to the underprivileged in society, the young, the dispossessed. He
talked about the economic needs of the community, he spoke of their victorious past, and he promised a future for them. For the first time the Muslims
outside the closed circle of the elite were being addressed. He was probably the
first All-India Muslim leader who was specifically referring to economic issues,
as Iqbal had advised him. (Ahmed 1997, 76)

Another interesting comparison between the two is that both Jefferson
and Jinnah were lawyers and in love with the idea that the Constitution,
which reflects the finest of human reason and civilization, can guide citizens and ensure their political and religious rights. Both their nations
were born in situations of revolution, murder, and mayhem. Both nations
were born with great hopes.

Both men would emphasize religious and political freedom and equally
emphasize education. Jefferson's love of education helped him to create the
University of Virginia which still honors his name. He always maintained his
affection for the College of William and Mary which he entered in 1760.
After leaving the presidency he spent his energy in supporting the university.
Jinnah too would keep in close touch with several educational institutions
including the Sind Madrassah in Karachi where he was educated. He had a
close relationship with Aligarh University and the students from Aligarh
would act as his most loyal troops in the battle for Pakistan. When his will
was disclosed people were amazed to see that he had left his fortune to the
Sind Madrassah, Aligarh, and colleges in Delhi and Bombay. They were
amazed because Delhi, Aligarh, and Bombay were not in Pakistan but in
India. Jinnah never changed his will even after he left India for Pakistan.

Both were born as subjects of the British Empire; both ended their careers
by leading a successful revolution against the British. Both were religious in
a general, broad sense and not dogmatic or what today would be called fundamentalist. Both had little time for the obsessive minutiae of religion while
respecting the moral strength that religious tradition provides. The two icons
of their nations would have found much in common with each other.

For Jefferson free public education was important to create a democratic
nation: "If a nation expects to be ignorant and free it expects what never was
and never will be." Education was the antidote to political tyranny:
"Enlighten the people generally and tyranny and oppression of body and
mind will vanish like spirits at the dawn of day."

The American Declaration of Independence is one of the most eloquently
written political texts in history. The famous sentence that talks of the
immutable, inalienable, and natural laws that declare certain truths as "self-
evident" is one of the most famous in American political mythology. To
Jefferson these truths are "sacred and undeniable." For him they were part
of the very social order and therefore "inherent." The tone echoes the high

moral vision that is inspired by a spiritual understanding of the purpose of existence. Jefferson's contribution to the Declaration and his writing elsewhere make him the preeminent interpreter of America's vision of itself.

Jinnah also left behind a memorable and eloquent testament. His first speech to the first gathering of the Constituent Assembly of Pakistan in August 1947 contains a powerful message to the nation. He points out that "color, cast or creed" will not determine the rights of the citizen. All ethnic groups and all religions had equal rights. He ends on a powerful note:

> You are free; you are free to go to your temples, you are free to go to your mosques; or to any other place of worship in this State of Pakistan. . . . We are starting in the days when there is no discrimination, no distinction between one community and another, no discrimination between one cast or creed and another. We are starting with this fundamental principle that we are all citizens and equal citizens of one state.

Jinnah had great hopes for his nation. He pledged his own role:

> My guiding principle will be justice and complete impartiality, and I am sure that with your support and cooperation, I can look forward to Pakistan becoming one of the greatest nations of the world.

In spite of the similarities we can ask the legitimate question: Why did democracy fail to take root in Pakistan? Is it, as the media after September 11 ask, that Islam and democracy are incompatible?

The answer comes more from the realm of sociology than theology. While Islam undoubtedly underlines the egalitarian nature of society and therefore prepares the foundations for a democratic order, society in Pakistan was dominated from the start by feudal lords and tribal chiefs. Add to this the general illiteracy of the population—even now as high as about 70 percent—and the poor development of democracy is explained. Jinnah's early death one year after the creation of Pakistan further discouraged the development of democracy and his ideas. With the weakened democratic process, the one strong and organized force, the Pakistan army, cast covetous eyes on the offices of power. Within ten years of Jinnah's death, General Ayub Khan declared the first Martial Law in Pakistan. It would last a full decade and at the end of it disaster would follow. Pakistan would be broken in two and lose its eastern wing—which became Bangladesh.

Since September, worshippers in churches and mosques have been brutally killed. Danny Pearl, the Wall Street journalist, was savagely executed because his killers said he was Jewish. Doctors belonging to the Shia community have been assassinated by the dozen. Pakistan is in crisis. More than ever before it needs to rediscover the ideals of its founding father.

Consider their achievements. Jefferson was one of the founding fathers of the nation which within two centuries of its birth would become the sole superpower of the entire planet. Whatever it did or did not do would matter throughout the world. It was capable of great acts of charity and aid. It also attracted hostility and hatred. The targets for the hijackers on September 11 were symbolic: the financial heart represented by the twin towers in New York and military might symbolized by the Pentagon.

Jinnah's Pakistan was significant on several levels. By coming into existence in 1947 it became the largest Muslim nation on earth. Its span allowed it to play a key role in the era of the Cold War. It was the one nation allied to the United States with membership in both CENTO and SEATO. Today it remains important because of its location, its population, which is about half that of the United States, and its nuclear power. Equally important, Pakistan represented an ideology of a moderate Muslim nation based on the ideals of liberty and freedom as enunciated by its founder. It is a leading role model, potentially, for the Muslim world.

CONTROVERSY

No man in public life is free of controversy. Not even a genuine founding father. Both Jefferson and Jinnah continue to rouse debate and controversy well after their deaths. In the age of the media and its tendency to employ the method of the inquisition against celebrity figures, Jefferson and Jinnah provide ready-made targets. Films, documentaries, and publications support them or attack them in equal measure. The controversies keep their memory alive.

To make matters worse, scholars have adopted the unfortunate tendency to impose their frame of thinking onto the historical past. This means that the prejudices, values, and ideas of today are employed as a standard to judge people in the past. Thus Jefferson is dragged over the coals for his affair with Sally Hemings:

> For many of his countrymen, Thomas Jefferson has become an unlikely authority to invoke in behalf of a modern conception of human rights. The embarrassing contradiction between his abstract commitment to equality and his self-indulgent life as the slave-owning master of Monticello has become too gross to bear. The strong likelihood that he fathered one or more enslaved children by Sally Hemings—his paramour, consort, or mistress, as well as half-sister-in-law—has only compounded the sin. (Rakove 2002, 1)

Commentators have used Jefferson's support of the French Revolution to condemn him to an absurd degree by comparing him with apologists of Stalin and Mao:

> In the hands of Conor Cruise O'Brien, the distinguished diplomat and political commentator, Jefferson's glib rationalizations for the violence of the French Revolution foreshadow the naivete with which so many intellectuals condoned the murderous purges of Stalin or Mao, while his enthusiasm for states' rights places him closer to the rabidly anti-nationalist ideology of the so-called militia movement than to the new political order that most Americans have accepted since the 1930s. (O'Brien 1996)

There is little doubt that Jefferson, who had a general respect for humanity, nonetheless reflected the prejudices of his time. This prejudice cannot be condoned. To Jefferson the African-American reeks of "a very strong and disagreeable odor" (Jefferson 1984, 265). He is inclined towards sexual excitement and not "tender"; "more of sensation than reflection" (p. 265). "In reason much inferior" (p. 266). He cannot comprehend the Greeks: for example, Euclid. "In imagination they are dull, tasteless and anomalous" (p. 266).

Jefferson concedes that the African-Americans are gifted in music but points out that they cannot create poetry. He compares them to the slaves during the time of the Romans and the Greeks. There was a difference, he observes. The slaves of that time were white and therefore could appreciate art and literature. He concludes: "The blacks . . . are inferior to the whites in the endowments both of body and mind" (p. 270).

The question of the slaves created serious problems in the thinking of Jefferson. We cannot for a moment condone his ideas on racial differences. He himself hoped that change would eventually create a more harmonious society and suggested that the slaves be moved to their own country, which he did not specify. This change he hoped would come not through violence, but "with the consent of the masters, rather than by their extirpation" (p. 289).

The criticism of Jinnah is as complex as that received by Jefferson. Contemporary Muslim scholars criticize Jinnah for not being "Islamic" enough:

> Jinnah's modern personality incurs the wrath of Muslims like the late Kalim Siddiqui, who styled himself the leader of the British Muslims. Siddiqui described Jinnah as one of the "stooges of imperialism" (*Guardian*, 9 May 1992), and the extremist organization, the Hizb-ut-Tahrir, call him the "Imperialist Collaborator" on their posters. For them Jinnah was too influenced by the West and not Islamic enough. The fact that he created the largest Muslim nation on earth is ignored. The Hizb-ut-Tahrir demand a pure theocratic Islamic state in opposition to a corrupt and decadent West. Living in the UK as permanent citizens, they have antagonized many by demanding Islamic law in Britain including the chopping off of hands as punishment for thieves and calling for the veiling of women.

> "Jinnah defies Allah" screamed the subtitle of a special feature on Jinnah, "Mohammed Ali Jinnah exposed!", in the December 1996 issue of the Hizb magazine *Khilafah*. Its main argument was that Jinnah had been influenced by the *kafir* because he insisted on democracy and indeed went as far as saying that Islam stood for democracy. They condemned and quoted Jinnah's statements: "Islam believes in democracy" (speech at Aligarh Muslim University on 6 March 1940) and "We learned democracy 1300 years ago" (presidential address, Muslim League session on 24 April 1943). The article indignantly claimed that "Jinnah went one step further than most traitors, the man also had the audacity to justify his Kufr actions by Islam." The article then went on: "How dare this man associate a Kufr concept such as democracy with our Prophet (saw)!" They further accused him of "inner secularism" because he had included a Hindu in Pakistan's first cabinet. They condemned the speech

of August 1947 in which he spoke of tolerance for the minorities. The article concludes: "These are only some of the examples where Jinnah clearly defied Allah (swt) and chose the way of the Kafir over the way of the Prophet (saw):" (Ahmed 1997, 198–99)

CONCLUSION

It is commonly said that Jefferson's ideas are a product of the European Enlightenment. The European Enlightenment, in turn, takes its inspiration from the Greeks. There is, of course, a connection between the Greeks and the Muslims living in Europe. When Muslims ruled the Iberian Peninsula a thousand years ago, they created a rich civilization based on the synthesis of cultures with Jewish, Christian, and Muslim traditions. Greek thought was translated freely into Arabic and from Arabic, introduced to the rest of Europe. This infusion of Greek thinking laid the foundations of what would become the European Renaissance in the following few centuries. Without Arabic, Europe may never have discovered Aristotle, Socrates, and Plato— and Jefferson may well have been deprived of the Enlightenment.

Jinnah, too, may be said to be a product of the European Enlightenment, considering how impressed he was by first Lincoln's Inn and Westminster. However, Jinnah was constantly looking beyond the European Enlightenment, not to the Greeks but to the origins of Islam. Both men, inspired by two different sources, had arrived at the conclusion that human society was best organized and conducted when individual liberty, religious freedom, and universal education were guaranteed. In short, the study of Jefferson and Jinnah is more than a study of two founding fathers. It leads us to the startling conclusion that the two leaders, one Christian and one Muslim, illustrate for us the point of contact between Western and Islamic civilization. We need to be grateful to Jefferson and Jinnah for leading us to this unexpected but vitally important conclusion.

The argument above becomes vitally important in understanding the nature of our contemporary world. For those who talk of the clash of civilizations we can illustrate through a discussion of Jefferson and Jinnah that a dialogue of civilizations is not only desirable but also possible.

When talking to America, Muslim military dictators and royal dynasts argue that if America does not support them then the only alternative is the Taliban kind of leadership awaiting them. America can answer by pointing to the model of Mr. Jinnah. Similarly, when Muslim leaders see Americans as bullying imperialists bent on a war against Islam, Americans can read them the works of Jefferson, which support the idea of universal liberty and freedom.

The comparison between Jefferson and Jinnah is also a useful corrective in the interaction between the two cultures. Far too few Muslims, even those living in America, know much about Jefferson. In turn, Americans have little idea of Jinnah. It is therefore a challenge to introduce each to the culture of the other.

Jefferson continues to inspire. Bill Clinton's middle name is Jefferson in honor of this founding father of America. Unfortunately Jinnah still remains an aloof figure in Pakistani culture. Few Pakistanis call their children Jinnah. Ten thousand fathers named their sons Osama. The battle between those who want Jinnah's vision of Pakistan and those who advocate the ideas of Osama will influence the history of the twenty-first century and not only in South Asia.

To the questions raised at the start of the essay, we can answer in the affirmative: Yes, Islam is clearly compatible with democracy. Yes, there are democratic leaders in Islam. And yes, there can be a genuine meeting of minds between Islam and the West.

I believe the broad and general issues raised in this paper need to be further explored. This essay serves to point to some crucial areas of discourse in the ongoing global interaction between Western and Islamic civilizations. It is merely exploratory, suggesting ways forward in the hope of generating a dialogue of civilizations in the midst of the talk of the clash of civilizations.

BIBLIOGRAPHY

Afzal, M. Rafique, ed. (1966). *Speeches and Statements of the Quaid-i-Azam Mohammad Ali Jinnah, 1911–36 and 1947–48*. Lahore.

Ahmad, Jamil-ud-din, ed. (1976). *Speeches and Writings of Mr. Jinnah*. Lahore: Sh. Muhammed Ashraf.

Ahmad, Riaz (1994). *Quaid-i-Azam Mohammad Ali Jinnah: Second Phase of His Freedom Struggle 1924–34*. Islamabad: Quaid-i-Azam University.

Ahmed, Rizwan (1993). *"Sayings of Quaid-i-Azam": Mohammed Ali Jinnah*. Karachi: Quaid Foundation and Pakistan Movement Centre.

Ahmad, Ziauddin (1970). *Mohammad Ali Jinnah: Founder of Pakistan*. Karachi: Ministry of Information and Broadcasting.

Ahmed, Akbar S. (1994). "Jinnah and the Quest for Muslim Identity." *History Today* 44 (September).

Ahmed, Akbar S. (1997). *Jinnah, Pakistan and Islamic Identity: The Search for Saladin*. London and New York: Routledge.

Albanese, Catherine L. (1976). "The Civil Religion of the American Revolution." *Sins of the Fathers*. Philadelphia: Temple University Press.

Bolitho, Hector (1954). *Jinnah: Creator of Pakistan*. London: John Murray.

Breen, T.H. (2001). "The Lockean Moment: The Language of Rights on the Eve of the American Revolution." Inaugural Lecture delivered before the University of Oxford, Oxford University Press, May 15.

Buckley, Thomas E.S.J. (1977). *Church and State in Revolutionary Virginia, 1776–1787*. Charlottesville: University Press of Virginia.

Commager, Henry Steele (1977). *The Empire of Reason: How Europe Imagined and America Realized the Enlightenment*. Garden City, New York: Doubleday.

Cunningham, Noble E. Jr. (1987). *In Pursuit of Reason: The Life of Thomas Jefferson*. Baton Rouge: Louisiana State University Press.

Elshtain, Jean Bethke (1995). *Democracy on Trial*. New York: Basic Books.

Gaustad, Edwin S. (1996). *Sworn on the Altar of God: A Religious Biography of Thomas Jefferson*. Grand Rapids, Michigan and Cambridge, UK: William B. Eerdmans Publishing Company.

Hasan, Syed Shamsul (1976). . . . *Plain Mr. Jinnah*. Karachi: Royal Book Company.

Jefferson, Thomas (1984). *Thomas Jefferson: Writings*. Ed. Merrill D. Peterson. New York: The Library of America, distributed by Penguin Putnam Inc.

Lerner, Max (1996). *Thomas Jefferson: America's Philosopher-King*. New Brunswick, New Jersey, USA and London, UK: Transaction Publishers.

Malone, Dumas (1948–81) *Jefferson and His Time*. Boston: Little, Brown.

May, Henry F. (1976). *The Enlightenment in America*. New York: Oxford University Press.

Mead, Sidney E. (1977). *The Old Religion in the Brave New World: Reflections on the Relation Between Christendom and the Republic*. Berkeley and Los Angeles: University of California Press.

Merchant, Liaquat H. (1990). *Jinnah: A Judicial Verdict*. Karachi: East & West Publishing Company.

Meyer, Donald H. (1976). *The Democratic Enlightenment*. New York: Putnam.

O'Brien, Conor Cruise (1996). *The Long Affair: Thomas Jefferson and the French Revolution, 1785–1800*. Chicago and London: University of Chicago Press.

Onuf, Peter S., ed. (1993). *Jeffersonian Legacies*. Charlottesville: University Press of Virginia.

Peterson, Merrill D. (1970). *Thomas Jefferson and the New Nation*. New York: Oxford University Press.

Peterson, Merrill D., ed. (1986). *Thomas Jefferson: A Reference Biography*. New York: Scribner's.

Pierard, Richard V. and Linder, Robert D. (1988). *Civil Religion and the Presidency*. Grand Rapids: Academie Books.

Reichley, A. James (1985). *Religion in American Public Life*. Washington DC: The Brookings Institution.

Richey, Russell E. and Jones, Donald G., eds. (1974). *American Civil Religion*. New York: Harper & Row.

Rakove, Jack N. (2002). "Jefferson, Rights, and the Priority of Freedom of Conscience." Paper presented at the Bellagio Conference organized by the Jefferson Center, June.

Sanford, Charles B. (1984). *The Religious Life of Thomas Jefferson*. Charlottesville: University Press of Virginia.

Sheldon, Garrett Ward (1991). *The Political Philosophy of Thomas Jefferson*. Baltimore: Johns Hopkins University Press.

Sheldon, Garrett Ward (1998). *What Would Jefferson Say?* New York: Penguin.

Sheldon, Garrett Ward (2000). *Jefferson & Atatürk: Political Philosophies*. New York: Peter Lang Publishing Inc.,

Sher, Richard B. and Smitten, Jeffrey R., ed. (1990). *Scotland and America in the Age of the Enlightenment*. Princeton: Princeton University Press.

Sloan, Douglas (1971). *The Scottish Enlightenment and the American College Ideal*. New York: Teachers College Press.

Talbot, Ian (1984). "Jinnah and the Making of Pakistan." *History Today* 34, (February).

Trench, Charles Chenevix (1987). *Viceroy's Agent*. London: Jonathan Cape.

Wasti, Syed Razi (1996). *At Quaid's Service*. Lahore: Jinnah-Rafi Foundation.

Wasti, Syed Razi, ed. (1994). *My Dear Quaid-i-Azam (Jinnah-Rafi Correspondence)*. Lahore: Jinnah-Rafi Foundation.

Wills, Garry (1978). *Inventing America: Jefferson's Declaration of Independence*. New York: Doubleday, Garden City.

Wills, Garry (1990). *Under God: Religion and American Politics*. New York: Simon & Schuster.

Wolpert, Stanley (1984). *Jinnah of Pakistan*. New York: Oxford University Press.

Zaidi, Z.H. (1993). *Jinnah Papers: Prelude to Pakistan*. Vol. I, parts I and II. Islamabad: Quaid-i-Azam Papers Project.

Zaidi, Z.H., ed. (1976). *M. A. Jinnah: Ispahani Correspondence 1936–1948*. Karachi: Forward Publications Trust.

Ziegler, Philip (1995). "Mountbatten Revisited." Faculty Seminar on British Studies, Austin, TX: The University of Texas at Austin.

PART II

RIGHTS AND THE CRAFTING OF CONSTITUTIONS

CHAPTER 6

NEED FOR A CREDIBLE MECHANISM TO SECURE ACCOUNTABILITY

Justice J.S. Verma

ACCOUNTABILITY: A FACET OF THE RULE OF LAW

The UN Secretary General, Kofi Annan, observed "*expansion of the rule of law in international relations has been the foundation of much of the political, social and economic progress achieved in recent years.*" He expressed the hope that it would "*facilitate further progress in the new Millennium.*"[1] He said:

> Since the founding of the United Nations in 1945, over 500 multilateral treaties have been deposited with the Secretary General...The aspirations of nations and of individuals for a better world governed by clear and predictable rules agreed upon at the international level are reflected in these instruments. They constitute a comprehensive international legal framework covering the whole spectrum of human activity, including human rights, humanitarian affairs, the environment, disarmament, international criminal matters, narcotics, outer space, trade, commodities and transportation. The norms of international behaviour expressed through these treaties make the modern world a far better place to live in than before...It is my hope that, as we enter the twenty-first century, nations would leave behind a world that was governed for most of history by a reliance on might and become more dependent on the international rule of law as envisaged by the Charter to guide their relations among each other.[2]

Commenting on the International Criminal Tribunals for the former Yugoslavia and Rwanda (ICTY), Martha Minow[3] observed:

> To respond to mass atrocity with legal prosecutions is to embrace the rule of law. This common phrase combines several elements. First, there is a commitment to redress harms with the application of general, pre-existing norms.

Second, the rule of law calls for administration by a formal system itself committed to fairness and opportunities for individuals to be heard both in accusation and in defense. Further, a government proceeding under the rule of law aims to treat each individual person in light of particular demonstrated evidence. In the western liberal legal tradition, the rule of law also entails the presumption of innocence, litigation under the adversary system, and the ideal of government by laws, rather than by persons. **No one is above or outside the law**, and no one should be legally condemned or sanctioned outside legal procedures....

A trial in the aftermath of mass atrocity, then, should mark an effort between vengeance and forgiveness. It transfers the individuals' desire for revenge to the state or official bodies. The transfer cools vengeance into retribution, slows judgment with procedure, and interrupts, with documents, cross-examination, and the presumption of innocence, the vicious cycle of blame and feud. The trial itself steers clear of forgiveness, however. It announces a demand not only for accountability and acknowledgement of harms done, but also for unflinching punishment. At the end of the trial process, after facts are found and convictions are secured, there might be forgiveness of a legal sort: a suspended sentence, or executive pardon, or clemency in light of humanitarian concerns. Even then, the process has exacted time and agony from, and rendered a kind of punishment for defendants, while also accomplishing change in their relationships to prosecutors, witnesses, and viewing public...

A tribunal can be but one step in a process seeking to ensure peace, to make those in power responsible to law, and to condemn aggression.

I illustrate my points with the help of two examples from the Indian scenario. In *Vineet Narain v. Union of India*,[4] the Supreme Court of India considered several issues of considerable significance to the rule of law. They began as yet another complaint of inertia by the Central Bureau of Investigation (CBI) in matters where the accusations made were against high dignitaries. The primary question was: "Is it within the domain of judicial review and could it be an effective instrument for activating the investigative process that is under the control of the executive?" The focus was on the question of whether any judicial remedy is available in such a situation. However, as the case progressed, it required innovation of a procedure christened 'continuing mandamus' within the constitutional scheme of judicial review, to permit intervention by the court to find a solution to the problem. This case has helped to develop a procedure within the discipline of law for the conduct of such a proceeding in similar situations. It has also generated awareness of the need of probity in public life and provides a mode of enforcement of accountability in this sphere. The Supreme Court took that opportunity to deal with the structure, constitution, and permanent measures necessary for having a fair and impartial agency. The faith and commitment to the rule of law exhibited by all concerned in these proceedings is the surest guarantee of the survival of democracy, of which rule of law is the bedrock. The basic postulate of the concept of equality: "Be you ever so high, the law is above you," governed all steps taken by the Supreme Court of India in these proceedings.

In recent weeks, the National Human Rights Commission of India (NHRCI) has been much preoccupied with events that have occurred in the State of Gujarat beginning with the Godhra tragedy on February 27, 2002 and continuing with the violence that ensued subsequently, in respect of which both the president and prime minister of India have expressed their deepest anguish. A team of the Commission visited Gujarat between March 19 and 22, 2002 and, in a series of Proceedings, the Commission has continued to monitor the situation closely. On April 1, 2002, the Commission made public its Preliminary Comments and Recommendations on the situation holding, inter alia,

...it is the primary and inescapable responsibility of the State to protect the right to life, liberty, equality and dignity of all of those who constitute it. It is also the responsibility of the State to ensure that such rights are not violated either through overt acts, or through abetment or negligence.

The Commission added

...it is a clear and emerging principle of human rights jurisprudence that the State is responsible not only for the acts of its own agents, but also for the acts of non-State players acting within its jurisdiction. The State is, in addition, responsible for any inaction that may cause or facilitate the violation of human rights.

A number of precise recommendations have been made by the Commission to bring to justice those responsible for the violations of rights that have occurred and to ameliorate the suffering of those who are the victims. The Proceedings of the Commission have been sent to the central and state governments for their response, after receiving which the Commission will reflect on what further needs to be said and done. The full text of the Opinion may be read on our website (http://nhrc.nic.in). The Gujarat example illustrates the role of independent national human rights institutions in enforcing accountability. Though protection of human rights is the primary responsibility of the judiciary, the NHRCI and the Indian judiciary have worked in ways complementing each other in securing and enforcing accountability. A similar complementarity could develop between corresponding institutions constituting the mechanism to enforce accountability at the international level.

Past Attempts in Securing Accountability

Except in the case of a total defeat or subjugation—for example, Germany after World War II—prosecutions of enemy personnel accused of war crimes have been both rare and difficult. National prosecutions have also been rare because of nationalistic, patriotic, or propagandistic considerations. The failure to prosecute some Germans who were accused of war crimes during World War I is an apt example. On the other hand, after the four principal victorious and occupying powers established an International Military Tribunal (IMT) following World War II, several thousand Nazi war criminals were tried either by national courts under Allied Control Council Law No. 10 or by various states under national decrees. Nuremberg's IMT, before which about 20 major offenders were tried, and the national courts, functioned reasonably well; the Allies had supreme authority over Germany and thus could often find and arrest the accused, obtain evidence and make arrangements for extradition. Of course, these trials attracted the criticism of "victor's justice."

In September 1946, the IMT at Nuremberg in its Judgment observed

Crimes against international law are committed by men, not by abstract entities, and only by punishing individuals who commit such crimes can the provisions of international law be enforced.

In the following years, the concept of individual criminal responsibility has taken root in the international law. Scholars say

*more than a half-century later, the response of the world's governments to this car-
dinal challenge seems lukewarm. International law has only haltingly and incom-
pletely recognized individual responsibility for human rights abuses in peace and
war, and states have only established or engaged domestic or international fora to
hold persons accountable sporadically and often with reluctance.*

They also note that the present legal environment resembles more a

*patchwork than a coherent, let alone complete, regime—criminal liability under
treaty law overlapping with some criminality under customary law; general rules
common to most systems of law that expand or contract the culpability of an
offender; a variety of rights and obligations to prosecute, but often with no will or
ability to do so; a few ad hoc international tribunals and an International
Criminal Court with limited jurisdiction [There is no agreement yet on the crime
against aggression; the terrorist and drug trafficking crimes excluded]; and special procedures that states have legislated or invoked to provide a non-prosecutorial
form of accountability.*[5]

Chile adopted a strategy for dealing with its past by granting amnesty for crimes against humanity. Several other countries also adopted a similar approach while dealing with atrocities committed in the transition from oppressive rule to democratic rule. South Africa (Truth and Reconciliation Commission), Guatemala, El Salvador, and Haiti are some recent examples.

Commenting on the failure to prosecute war crimes during the Gulf War, Theodor Meron says

The Persian Gulf War, as an international war, provided a classic environment for
the vindication of the laws of war so grossly violated by Iraq by its plunder of
Kuwait, its barbaric treatment of Kuwait's civilian population, its mistreatment

of Kuwaiti and allied prisoners of war and during the sad chapter of the U.S. and other hostages. Although the Security Council had invoked the threat of prosecutions of Iraqi violators of international humanitarian law, the ceasefire resolution did not contain a single word regarding criminal responsibility. Instead, the U.N. Resolution promulgated a system of war reparations and established numerous obligations for Iraq in areas ranging from disarmament to boundary demarcation.

This result is not surprising, for the U.N. coalition's war objectives were limited, and there was an obvious tension between negotiating a ceasefire with Saddam Hussein and demanding his arrest and trial as a war criminal. A historic opportunity was missed to breathe new life into the critically important concept of individual criminal responsibility for the violations of the laws of war. At the very least, the Security Council should have issued a warning that Saddam and other responsible Iraqis would be subject to arrest and prosecution under the grave breaches of provisions of the Geneva conventions whenever they set foot abroad.[6]

In response to serious violations of humanitarian law in the former Yugoslavia and Rwanda, the Security Council set up two ad hoc tribunals. Despite their other limitations, they also give rise to allegations of "selective justice." It can be argued with justification as to why those tribunals could not be established to deal with similar violations noticed in Vietnam, Cambodia, Burundi, Sudan, and many other places.

NEED FOR A PERMANENT MECHANISM

Experts say that the past experiences[7] of ad hoc international tribunals confirm the need for a permanent system of international criminal justice. Further, it is said

> A permanent system would eliminate the necessity of establishing ad hoc tribunals every time the need arises. The decision to establish such tribunals, not to mention drafting the applicable statutes, takes considerable time during which the evidence of the crimes becomes more difficult to obtain, and the political will to prosecute dissipates. Moreover, a political debate is invariably reopened over the provisions of the statute, who will conduct the prosecutions, and who will sit in judgment. Such pressures leave ad hoc tribunals vulnerable to political manipulation.[8]

As the evidence of human rights violations is either scarce or rapidly disappearing, there is a need to apprehend violators in double quick time, which is presently a challenge faced by the ICTY. Further, there are other challenges of lack of funding, personnel, and world interest.

Critics say that the major barrier to the creation of an International Criminal Court (ICC) in the past has been

> ...the conflict between state sovereignty and the jurisdiction of such a court. States are generally reluctant to expose their citizens (especially politicians and

senior military commanders) to potential criminal prosecutions for conduct undertaken in the name of the state.[9]

Scholars point out

the field of criminal law is seen to be closely associated with state sovereignty, with the ultimate application of the power of the state to persons within its territory or jurisdiction. National courts routinely apply foreign private or civil law, but the criminal law is seen as a matter of local public policy, a matter for the lex fori.[10]

Commenting on the deterrent effect of the ICTY, Theodor Meron observed in 1993 as follows:

...First, modern media ensures that all actors in the former Yugoslavia know of the steps being taken to establish the tribunal. Second, the tribunal will probably be established while the war is still being waged. Even the worst war criminals involved in the present conflict know that their countries will eventually want to emerge from isolation and be reintegrated into the international community. Moreover, they themselves will want to travel abroad. Normalization of relations and travel would depend on compliance with warrants of arrest.

His words have indeed proved very prophetic. Despite the issue of arrest warrant, Slobodan Milosevic was not handed over to the ICTY for a long time, but eventually the need for funds to reconstruct the former Yugoslavia and international pressure led to his surrender to the ICTY. The current publicity of his trial in the media will certainly instill fear in the minds of future dictators.

Commenting on the ICTY, Henry Steiner and Philip Alston[11] note the following:

The ICT is in a radically different situation from a court in a state observing fundamental principles of the Rule of Law in the sense that the state's executive and legislative branches comply with and execute court judgments. The Security Council has created an independent organ, as must be the case. Nonetheless, the ICT remains dependent on an uncertain and changing political context; it lacks the relative autonomy of a court in a state with a strong tradition of an independent judiciary. The Tribunal depends for funds on a UN General Assembly whose members hold different views about it and who may judge its work differently. It must receive support from states and from the Security Council with respect to such basic matters as putting pressure on states to comply with its orders. There is no equivalent to a 'national tradition' for the Tribunal to draw on.

The above holds true for the ICC as well, since its performance will eventually hinge on state cooperation and funding.

Though governments are yet to evolve a comprehensive and an authoritative international criminal code, international law clearly makes individuals responsible for certain gross violations of human dignity during peace and war, namely, genocide, crimes against humanity, war crimes, slavery and forced labour, torture, apartheid, and forced disappearances. The above international crimes have been proscribed in treaties and customary law. The Rome Statute of the ICC, 1998 also defined and criminalized several of these crimes. *Treaties and custom have, moreover, adopted various strategies for imputing individual responsibility, most notably by declaring an act to be an international crime or by obligating states to extradite offenders or punish them under different jurisdictional bases.*[12]

An international crime is presumably subject to universal jurisdiction. Piracy was regarded as an offence against "the law of nations." In the same vein, genocide, war crimes, unlawful seizing of aircraft, taking of hostages, and torture are subject to universal jurisdiction in the customary international law.

In the decades following the Nuremberg Judgment, there is a greater clarity in the crimes which constitute "war crimes" and "crimes against humanity." Their codification was an important development. The standard setting in human rights in the last few decades by way of conclusion of new treaties that address specific themes has been very impressive. The International Humanitarian Law, of course, evolved on a separate track. What is, however, regrettable is the lack of a credible enforcement mechanism. Alongside the evolution of human rights norms and treaties, massive human rights violations have also occurred. The World has been a witness to egregious violations during the Holocaust, in Bosnia, Rwanda, the apartheid regime of South Africa, early years of the Pinochet government in Chile, to mention but a few instances.[13] Religion, race, and ethnic tradition were some of the underlying components in these massive tragedies. They also involved savage dehumanization and hatred, often stimulated by an oppressor state. In many of the above instances, the perpetrators have gone unpunished, which only contributes to impunity.[14]

The principle of accountability requires that crimes do not and shall not go unpunished and that leaders are held to high standards of civilized behavior. The idea of accountability for massive human rights violations is highly pertinent to our times. Throughout recorded history, most egregious crimes—the mass atrocities, the genocides, and the crimes against humanity—have been ordered by leaders and it is for this reason they must be held primarily accountable.

Though important humanitarian law and human rights standards have evolved during the last one hundred years or so, the absence of a permanent enforcement mechanism was felt for long. As a result, perpetrators of some of the worst human rights violations have gone unpunished. Though international humanitarian law and human rights law gained wide acceptance, it was seldom enforced. Hans Corell[15] in fact termed the ICC a "missing link."

On April 11, 2002, the sixtieth ratification of the Rome Statute was deposited with the United Nations, paving the way for the creation of the world's first permanent ICC. Hailing this event, the UN High Commissioner for Human Rights, Ms. Mary Robinson said

> The unequivocal message emerging from The Hague and Arusha is that where domestic legal order has broken down, or national authorities are unwilling or unable to punish gross violations and abuses of human rights and international humanitarian law, the international community has an obligation and a responsibility to respond...With the coming into force of the Rome Statute, the international community will have accepted that responsibility on a permanent basis.

In his reaction, the UN secretary general, Kofi Annan said

> Impunity has been dealt a decisive blow. The time is at least coming when humanity no longer has to bear impotent witness to the worst atrocities, because those tempted to commit such crimes will know that justice awaits them.

The president of the UN General Assembly, Han Seung-soo of the Republic of Korea noted that the Court would limit the extent or duration of violence by the nature of its existence and that it will provide much stronger deterrence to potential criminals by giving them a clear warning that there will be no place for them to hide. However, the primary responsibility for the punishment of crimes is with the states and the ICC would step in only if the national system is unable or unwilling to do so.

The *jus cogens* nature of the international crime of torture makes it imperative for states to take universal jurisdiction over torture wherever committed. International law provides that offences *jus cogens* may be punished by any state because the offenders are "common enemies of all mankind and all nations have an equal interest in their apprehension and prosecution....As there is no tribunal or court to punish the international crime of torture, local courts could take jurisdiction. The objective is to ensure a general jurisdiction so that the torturer is not safe wherever he went. The Torture Convention provides an international system under which the international criminal—torturer—could find no safe haven."[16]

Lord Browne-Wilkinson, a member of the Appellate Committee to examine the principal charge of torture against Pinochet, observed as follows:

> Can it be said that the commission of a crime which is an international crime against humanity and jus cogens is an act done in an official capacity on behalf of the state? I believe there to be strong ground for saying that the implementation of torture as defined by the Torture Convention cannot be a state function. This is the view taken by Sir Arthur Watts ['The Legal Position in International Law of Heads of States, Heads of Government and Foreign Ministers'] who said (at p. 82)
> "While generally international law...does not directly involve obligations on individuals personally, that is not always appropriate, particularly for acts of such seriousness that they constitute not merely international wrongs (in the broad sense of a civil wrong) but rather international crimes which offend

against the public order of the international community. States are artificial legal persons: they can only act through the institutions and agencies of the state, which means, ultimately through its officials and other individuals acting on behalf of the state. For international conduct which is so serious as to be tainted with criminality to be regarded as attributable only to the impersonal state and not to the individuals who ordered or perpetrated it is both unrealistic and offensive to common notions of justice."

It can no longer be doubted that as a matter of general customary international law a head of state will personally be liable to be called to account if there is sufficient evidence that he authorized or perpetrated such serious international crimes.[17]

Commenting on this observation of Sir Arthur, Lord Goff of Chieveley in the Pinochet case said

It is evident from this passage that Sir Arthur is referring not just to a specific crime as such, but to a crime which offends against the public order of the international community, for which a head of state may be *internationally* [his emphasis] accountable. The instruments cited by him show that he is concerned here with crimes against peace, war crimes and crimes against humanity. Originally these were limited to crimes committed in the context of armed conflict, as in the case of the Nuremberg and Tokyo Charters, and still in the case of the Yugoslavia Statute... Subsequently, the context has been widened to include (inter alia) torture "when committed as part of a widespread or systematic attack against a civilian population" on specific grounds. A provision to this effect appeared in the International Law Commission's Draft Code of Crimes of 1996... and also appeared in the Statute of the International Tribunal for Rwanda (1994) and in the Rome Statute of the International Court (adopted in 1998).... These instruments are all concerned with international responsibility before international tribunals, and not with the exclusion of state immunity in criminal proceedings before national courts...[18]

Lord Millett in the Pinochet case noted as follows:

The case [The opinion considered in the "land mark decision" of the Supreme Court of Israel in Attorney General of Israel v. Eichmann] is authority for three propositions:

1. There is no rule of international law which prohibits a state from exercising extraterritorial criminal jurisdiction in respect of crimes committed by foreign nationals abroad.
2. War crimes and atrocities of the scale and international character of the Holocaust are crimes of universal jurisdiction under customary international law.
3. The fact that the accused committed the crimes in question in the course of his official duties as a responsible officer of the state and in the exercise of his authority as an organ of the state is no bar to the exercise of the jurisdiction of a national court.

...crimes prohibited by international law attract universal jurisdiction under customary international law if two criteria are satisfied. First, they must be contrary to a peremptory norm of international law so as to infringe a jus cogens. Secondly, they must be so serious and on such a scale that they can justly be regarded as an attack on the international legal order. Isolated offences, even if committed by public officials, would not satisfy these criteria...[19]

IMPLICATIONS OF THE PINOCHET DECISION

Izzat Ibrahim al-Duri, the number-two man in Iraq after Saddam Hussein, who was in Vienna, Austria for medical treatment, had to make a hasty exit as a criminal complaint was filed with Austrian authorities over the mass murder of Kurds in 1988 and torture and killing of Iraqi citizens. Pinochet Syndrome constrained the former president of Indonesia, Mr. Suharto, from traveling to Germany for medical treatment. International nongovernmental organizations have compiled a list of ex-tyrants who have fled their countries for what they thought were safer addresses. They include Idi Amin of Uganda who is still in Saudi Arabia; Jean-Claude Duvalier of Haiti who is in France; one of his successors, Raul Cedras in Panama; and Paraguay's Alfredo Stroessner in Brazil. Drawing on the legal precedent established in Pinochet case, human rights groups have documented violations committed by the former president of Chad, Hissene Habre, who has been living in exile in Dakar since being toppled from power in 1990. He has been in Dakar under house arrest since being indicted on torture charges by the court. "*This is a message to other African leaders that nothing will be the same any longer,*" said Delphine Djiraibe, president of a human rights group working on the case.[20]

The decisions in Pinochet 1 and Pinochet 3, which denied the senator sovereign immunity, have more obvious international implications. The rulings provide a precedent for limiting claims of immunity by former heads of state and thus open the way for their more general prosecution.

Can we leave it to individual states to become the human rights police of the world? There are some who argue about violations committed by leaders while promoting humanitarian causes as in NATO bombing in Serbia.[21]

While upholding the common view that there exists universal jurisdiction for crimes against humanity, by which states can try in their national courts persons alleged to have committed crimes against humanity, Ben Chigara argues

the Pinochet case indicates, first, that a state which, for whatever reason, grants amnesty to persons, who would otherwise be charged with crimes against humanity, should register such an amnesty with the united Nations, to ensure the acceptance or acquiescence by other states of that decision; and secondly, that international crimes should at all times be tried by international tribunals. There is a need to evaluate the place in international law of national amnesties as a mechanism by which a state may choose to deal with an uncomfortable immediate past with the hope of securing a future that is entirely different and which marks a new stage in the nation's history.[22]

The two treaties most applicable to this case, the Convention against Torture and Other Cruel, Inhuman or Degrading Treatment or Punishment (1984), which came into force on June 26, 1987, and the International Convention Against the Taking of Hostages (1979), which came into force in 1983, both authorize states to adopt measures for their enforcement. By Article 2 of the Convention Against Torture, each state-party is obligated to take effective legislative, administrative, judicial, or other measures to prevent acts of torture in any territory under its jurisdiction. This same provision rules out, as a justification of torture, any exceptional circumstances whatsoever, whether a state of war or a threat of war, internal political instability or any other public emergency. Moreover, an order from a superior officer or a public authority may not be invoked as a justification of torture. More importantly, perhaps, for the purposes of this discussion, by Article 5(1) each state party shall take such measures as may be necessary to establish its jurisdiction over the offences referred to in Article 4 in cases: (a) when the offences are committed in any territory under its jurisdiction or on board a ship or aircraft registered in that state; (b) when the alleged offender is a national of that state; and (c) when the victim is a national of that state if that state considers it appropriate.[23]

Thus, the implications of the Pinochet case are far reaching, and there is need for settled norms to prevent its misuse by the powerful against the not so powerful, as also for its equal application.

SOME CHALLENGES

Issuing indictment is one thing. Arresting the indicted leaders and presenting them before an international tribunal for trial in the face of a non-cooperative government is quite another. If the tribunals do not have their way in securing the presence of indicted leaders, they become a farce and their credibility gets severely eroded. There will be loss of public confidence too. Yugoslavia held out for long despite repeated requests from the ICTY to hand over top leaders including Slobodan Milosevic. In such circumstances, the tribunals attract criticism that they can at best secure the arrest of lower-rung leaders involved in the atrocities while the big fish get away. As of now, there have been no instances of arrests of indicted leaders by an international force when confronted with opposition by governments to hand them over for trial.

Are we in a position today to assert that egregious human rights violations would be addressed squarely wherever they occur and by whomsoever they are committed? Is the international mechanism robust enough to overcome political considerations? What is the competence of the ICC vis-à-vis violations occurring in the countries of the five permanent Members of the Security Council? How can we rid the employment of double standards in the application of human rights standards in the relations between States? There are numerous instances where trade, economic, and strategic interests have forced human rights considerations to either take a back seat or be used for coercion, and how these considerations in a country's foreign policy have made it turn a blind eye to violations in a friendly regime. In other words, there ought to be greater consistency in the application of human rights standards. There is a need to ensure that political considerations do not come in the way of legal considerations.

CONCLUSION

There is a felt need for a credible international mechanism to enforce accountability, which must ensure swift trials and punishment for gross abuses, whenever and wherever they occur. Accountability can be secured only through a credible enforcement mechanism. The Rome Statute of the ICC will enter into force on July 1, 2002. Let us hope that it does not attract the criticism of becoming a political instrument of the powerful for policing the global village.

I would like to end by recalling the wise counsel of Mahatma Gandhi who gave a message even for this situation. He said

> Peace does not come out of a clash of arms,
> it comes from justice lived and done

An effective mechanism to administer justice by punishing equally all crimes against humanity alone can secure accountability for those crimes. The mechanism has to be equally efficacious against all offenders, irrespective of the power of the State to which they belong. Recent measures are a firm step in that direction but there is a long road yet to traverse. The need for a credible mechanism for doing justice for all wrongs can no longer be doubted.

NOTES

1. The secretary general's letter to heads of state and Government dated May 15, 2000, published in the "Millennium Summit Multilateral Treaty Framework: An Invitation to Universal Participation," United Nations, September 6–8, 2000.
2. Ibid, see foreword, ix–x.
3. Martha Minow, "Between Vengeance and Forgiveness," at 25, cited in Henry J. Steiner and Philip Alston, *International Human Rights in Context Law Politics Morals* (Clarendon Press, Oxford 1998), 1143–144.
4. 1998 (1) SCC 226. Consequent upon arrest and interrogation of an official of a terrorist organisation, raids were conducted by the Central Bureau of Investigation (CBI) on the premises of Surender Kumar Jain, his brothers, relatives, and businesses. Along with the Indian and foreign currency, the CBI seized two diaries and two notebooks from the premises. They contained detailed accounts of vast payments made to persons identified only by initials. The initials corresponded to the initials of various high-ranking politicians, in or out of power, and of high-ranking bureaucrats. Nothing having been done

in the matter of investigating the Jains or the contents of their diaries, the writ petitions were filed in public interest under Article 32 of the Constitution.

5. Steven R. Ratner and Jason S. Abrams, *Accountability for Human Rights atrocities in International Law Beyond the Nuremberg Legacy* (Oxford University Press, 2001).

6. Theodor Meron, "The Case for War Crimes Trials in Yugoslavia," 72 *Foreign Affairs* 122 no. 3 (1993) at 123.

7. ICTY could prosecute and convict only a few low-ranking Bosnian Serb violators, but is unable to obtain jurisdiction over scores of indicted persons, including major offenders, which IFOR refuses to apprehend. Rwandan tribunal's experience has not been very different with bureaucratic and financial difficulties. See Cherif Bassiouni, "From Versailles to Rwanda in Seventy-Five Years: The Need to Establish a Permanent International Criminal Court," *Harvard Human Rights Journal* 10 (1997), 58–61.

8. Ibid., 60–61.

9. See generally Peter Burns, "An International Criminal Tribunal: The Difficult Union of Principle and Politics," *Criminal Law Forum* 5, nos. 2–3 (1994), 341–80.

10. James Crawford, "The ILC adopts a Statute for an International Criminal Court," *The American Journal of International Law* 89 (1995), 404–16.

11. Steiner and Alston, *International Human Rights in Context Law Politics Morals*.

12. Ratner and Abrams, *Accountability for Human Rights atrocities*.

13. Pol Pot, Idi Amin, and Saddam Hussein have still not been brought to justice for the atrocities committed in Cambodia, Uganda, and against Kurds in Iraq respectively.

14. Steiner and Alston, *International Human Rights in Context Law Politics Morals*.

15. The UN Under-Secretary-General for Legal Affairs.

16. Diana Woodhouse, ed., *The Pinochet case, A Legal and Constitutional Analysis* (Hart Publishing, February 2000).

17. Steiner and Alston, *International Human Rights in Context Law Politics Morals*.

18. Ibid., 1207–209.

19. Ibid., 1210–212.

20. Ibid., 1214–216.

21. Jonathan Black-Branch, "Sovereign Immunity," in Diana Woodhouse, ed., *The Pinochet case, A Legal and Constitutional Analysis*, 112.

22. Ben Chigara, "Pinochet and the Administration of International Criminal Justice," in Diana Woodhouse, ed., *The Pinochet case, A Legal and Constitutional Analysis*.

23. Diana Woodhouse, ed., *The Pinochet case, A Legal and Constitutional Analysis*.

CHAPTER 7

RIGHTS AND HUMAN RIGHTS IN THE MODERN WORLD: THE EXPERIENCE OF WORKING THE BILL OF RIGHTS IN THE INDIAN CONSTITUTION

Soli J. Sorabjee

Part III of the Indian Constitution guarantees certain fundamental rights. It was not incorporated as a popular concession to international sentiment and thinking on human rights in vogue after the conclusion of the World War II. The demand for constitutional guarantees of human rights for Indians was made as far back as 1895 in the Constitution of India Bill, popularly called the Swaraj Bill. It was inspired by Lokmanya Tilak, one of the greatest freedom fighters and architects of India's independence. This Bill envisaged for India a Constitution guaranteeing to every one of its citizens freedom of expression, inviolability of one's house, right to property, equality before the law, equal opportunity of admission to public offices, right to present claims, petitions, and complaints and right to life and personal liberty.

Annie Besant's Commonwealth of India Bill, finalized by the National Convention of Political Parties in 1925 also embodied a specific declaration of rights visualizing for every person certain rights in terms practically identical with the relevant provisions of the Constitution of the Irish Free State in 1921.

The problem of minorities in India further strengthened the general demand in favor of inclusion of fundamental rights in the Indian Constitution. In the Madras Session 1927, the Indian National Congress laid down that the basis of a future Constitution must be a declaration of fundamental rights. In 1928, the Motilal Nehru Committee in its report recommended the adoption of these rights as a part of the future Constitution of India.

The Indian Statutory Commission, popularly known as the Simon Commission, reflected typical traditional British thinking when in 1930 it turned down the demand for fundamental rights on the ground that "abstract declarations are useless, unless there exist the will and the means to make them effective."

After the breakdown of talks between Gandhi and Jinnah in September 1944, a Non-Parties conference was convened. A committee was formed consisting of eminent persons and was headed by a legal luminary, Sir Tej Bahadur Sapru. The Sapru Committee was of the opinion that in the peculiar circumstances prevalent in India, fundamental rights were necessary not only as an assurance and guarantee to the minorities but also for prescribing a standard of conduct for the legislatures, governments, and the courts.

During the framing of the Constitution, the subject of fundamental rights was extensively debated in the Constituent Assembly. The debates in the Constituent Assembly reflect the belief of the founding fathers in the universality and inalienability of basic human rights.

Ultimately a comprehensive array of human rights was guaranteed in part III of the Indian Constitution. They broadly correspond to the International Covenant on Civil and Political Rights 1966 [ICCPR]. They comprise constitutional guarantees of equality, freedom of expression, assembly and association, freedom of movement, freedom to carry on profession and business, freedom of conscience and religion. There are guarantees against retrospective criminal laws, double jeopardy and self-incrimination, and against deprivation of life and personal liberty. There are constitutional provisions to prevent exploitation of children. Minorities are guaranteed linguistic and cultural rights, and the right to establish and administer educational institutions of their choice.

Fundamental rights are mainly enforceable against the State and its manifold instrumentalities as also against bodies and institutions in which there is significant government control, supervision, and involvement.

The philosophy underlying the Bill of Rights in the Indian Constitution is that a human being is not mere mass and molecules but there is a spiritual spark in every individual irrespective of race, religion, creed, caste, color, language, sex, or status. The rationale of human rights is that they flow from the common humanity, the inherent dignity of every human being, and the equal and inalienable rights of all members of the human family. It is a fallacy to regard a Bill of Rights as a gift from the State to its citizens. Individuals possess basic human rights independently of any Constitution by reason of the basic fact that they are members of the human race. A Bill of Rights does not "confer" fundamental human rights. It confirms their existence and accords them protection. Its purpose is "to withdraw certain subjects from the vicissitudes of political controversy, to place them beyond the reach of majorities and officials and to establish them as legal principles to be applied by the courts."

Part IV of the Constitution of India lays down Directive Principles of State policy. In substance they are in the nature of social and economic rights and have features which are common to the International Covenant on Social, Economic and Cultural Rights 1966 [ICSECR]. Although Directive Principles are not enforceable by any court, they are "nevertheless fundamental in the governance of the country and it shall be the duty of the State to apply these principles in making laws." The concept of Directive Principles was borrowed from the Constitution of Eire which also contains a set of Directive Principles.

Fundamental rights in the Indian Constitution are not merely ornamental. They are enforced against the State and its instrumentalities by an independent judiciary exercising power of judicial review. Legislative or executive action which is in breach of any fundamental right can be and has been invalidated.

A Bill of Rights is the conscience of the Constitution. An independent judiciary is its conscience keeper. A most generous Bill of Rights can be reduced to arid parchment promises by narrow and insensitive judicial interpretation. Consequently, a Constitution, and in particular that part of it which protects and entrenches fundamental rights and freedoms should be given a generous and purposive construction avoiding what has been called "the austerity of tabulated legalism," remembering that "the letter killeth, but the spirit giveth life."

The Indian judiciary has played a creative role in the interpretation of the Constitution. Fundamental rights that are not specifically mentioned have been spelt out and deduced on the theory that certain unarticulated rights are implicit in the enumerated guarantees.

Consider a few illustrations. The Constitution of India does not specifically guarantee freedom of the press as a fundamental right. In several decisions of the Supreme Court freedom of the press has been held to be implicit in the guarantee of freedom of speech and expression and has thus acquired the status of a fundamental right by judicial interpretation. There is no legislation in India as yet securing freedom of information. The Supreme Court by interpretation of the free speech guarantee deduced the right to know and the right of access to information on the reasoning that the concept of an open government is the direct emanation from the right to know, which is implicit in the guarantee of free speech and expression. The right to travel abroad has been spelt out from the expression "personal liberty" in Article 21 of the Constitution. Although there is no specific provision in the Constitution prohibiting cruel, inhuman, and degrading punishment or treatment, the Court has evolved this right from other provisions of the Constitution. The Court has deduced the right to elementary education up to the age of 14. Right to privacy has also been spelled out.

Similar exercises have been undertaken by the U.S. Supreme Court and Courts in the Irish Republic.

The expression "life" in Article 21 received an expansive interpretation. The Court ruled that "life" does not connote merely physical or animal existence but embraces something more, namely "the right to live with human dignity and all that goes along with it, namely, the bare necessaries of life such as adequate nutrition, clothing and shelter over the head."

Based on this interpretation the Supreme Court of India has ruled that the right to live with human dignity encompasses within its ambit the protection and preservation of an environment free from pollution of air and water. Health and sanitation have been held to be an integral facet of the right to life.

In its efforts to prevent environmental degradation the Court has ordered certain tanneries and chemical industries that were discharging effluents into the lakes and rivers to stop functioning, unless the effluents were subjected to a pre-treatment process by setting up primary treatment plants as approved by the State Pollution Boards. In its battle against pollution the Supreme Court has issued directions that all commercial vehicles in Delhi that were 15 years old and could cause vehicular pollution should be debarred from plying on public roads.

It is generally accepted that guaranteed fundamental rights are not absolute. They can be reasonably restricted in public interest. The question whether the restriction imposed is unreasonable, excessive, or disproportionate should be determined by an independent judiciary exercising the power of judicial review. This delicate judicial task of striking the balance requires understanding not merely of the legal and constitutional provisions but of the prevalent economic and sociological forces, the contemporary mores of society and a sense of history. The endeavor of Courts in India has been to achieve an acceptable accommodation of the conflicting interests of the individual, the society and the State. There is no royal road to attain such accommodation. Courts have on occasions not got the balance right. Perfection is not the attribute of common humanity, and judges have not been vouchsafed the divine gift of infallibility.

Certain core human rights are non-derogable or non-suspendable even during war or national emergencies. Under Article 4(2) of the ICCPR no derogation is permissible in respect of certain human rights namely, right to life; freedom from torture; prohibition against slavery and to be held in servitude; imprisonment merely on ground of inability to fulfil a contractual obligation; freedom from retrospective operation of criminal laws and double jeopardy; and right to freedom of thought, conscience, and religion.

In the Indian Constitution, as it stands at present, fundamental rights that are non-suspendable or non-derogable are protection of life and personal liberty; retrospective operation of criminal laws and double jeopardy; and protection against self-incrimination.

It may be of interest to note that the National Commission to Review the Working of the Constitution [NCRWC] in its Report submitted in March 2002 has recommended that some additional rights be made non-suspendable, namely prohibition of untouchability [Article 17]; prohibition of traffic in human beings and forced labor [Article 23]; prohibition of employment of children under the age of 14 in hazardous employment [Article 24]; freedom of conscience and religion [Article 25]; and the right to move the Supreme Court for enforcement of fundamental rights [Article 32].

The distinction between generational rights, namely civil and political liberties (first generation), social, economic, and cultural (second generation) and environmental (third generation) is a bit rigid. It fails to recognize the dynamic aspect of evolution of human rights. It would be more appropriate to regard the change in the idea of rights over a period of time as different "waves."

The first wave of human rights around the late eighteenth century, which witnessed the drafting of the American Bill of Rights and the French Declaration of the Rights of Man, was primarily concerned with guaranteeing liberty against state tyranny and from religious persecution. The second wave was generated by the massive atrocities committed by the Nazis before and during World War II. The present new wave of rights focuses upon the values of dignity, equality, and community. It has been aptly described as a search for certain basic values to guide human behavior. Dignity is the moral and intellectual source of human rights in present times.

The Vienna Declaration on Human Rights in June 1993 explicitly recognizes that "all human rights are universal, indivisible and interdependent and interrelated." This has put to rest the controversy regarding the superiority of one set of rights over the other. However, at the operational level in developing countries socio-economic rights would have priority in matter of implementation. For example, if the choice is between a new television tower, which would enhance freedom of expression, and the building of roads and hospitals limited financial resources would tilt the choice in favor of the latter.

The right to development has opened up new horizons in the field of human rights. The trend in recent national Constitutions is to give certain social, economic, and cultural rights the status of fundamental rights. For example, the Constitution of the Republic of South Africa 1996 includes in its Bill of Rights, the right to housing; health care; food, water, and social security; and education. The present Hungarian Constitution, in Article 70/E, contains social and economic rights framed in terms implying enforceability.

The most remarkable craftsmanship displayed by the Supreme Court of India in promoting human rights has been to incorporate into fundamental rights some of the Directive Principles, such as those imposing an obligation on the state to provide a decent standard of living, a minimum wage, just and humane conditions of work, and to raise the level of nutrition and of public health. This has been achieved by placing a generous interpretation on the expression "life" in Article 21 of the Constitution.

Access to justice is recognised as a basic human right. In order to achieve that it is necessary that the doctrine of *locus standi* should not be strict and rigid. A restrictive rule of *locus standi* in effect prevents protection and enforcement of rights of disadvantaged persons. The Supreme Court of India has liberalized this rule of standing in public law and ruled that where judicial redress is sought for legal injury done to indigent and disadvantaged persons, who on account of economic disabilities are unable to approach the courts themselves, any member of the public acting bona fide and not for oblique considerations, can maintain an action on their behalf.

Rights without remedies are useless. A mere declaration of invalidity of an executive order or an administrative decision which has resulted in the violation of a person's fundamental rights would not provide a meaningful remedy. In some Constitutions there is a provision for award of compensation to victims of unlawful arrest or detention. The ICCPR provides that "anyone who has been the victim of unlawful arrest or detention shall have an enforceable right to compensation" [see Article 9(6)]. The Indian Constitution contains no such provision. Nonetheless the Supreme Court has, in some cases, ordered compensation by the State as a remedy in public law. The NCRWC has recommended that right to compensation for violation of a person's life or liberty be made an enforceable fundamental right by an express provision in the Constitution.

Conception of human rights varies. As new thoughts and dimensions in the field of human rights emerge it may become necessary to amend the Constitution and specifically incorporate additional human rights in the Bill of Rights. No Constitution can be permanent. Therefore every Constitution has a provision for its amendment. Remember the wise words of one of the greatest architects of the American Constitution, Thomas Jefferson:

> Some men look at constitutions with sanctimonious reverence and deem them like the Ark of the Covenant too sacred to be touched. They ascribe to the men of the preceding age a wisdom more than human and suppose what they did to be beyond amendment. ... I am certainly not an advocate for frequent and untried changes in laws and constitution ... but I know that the laws and institutions must go hand in hand with the progress of human mind. ... As new discoveries are made, new truths discovered and manners and opinions change with the change of circumstances, institutions must advance also and keep pace with the times.

Pandit Nehru, the first prime minister of India, expressed the same thought in felicitous language.

> A constitution which is unchanging and static, it does not matter how good it is, how perfect it is, is a constitution that has past its use. It is in its old age already and gradually approaching its death. A constitution to be living must be growing; must be adaptable; must be flexible; must be changeable ... as society changes, as conditions change, we amend it in the proper way. It is not like the unalterable law of the Medes and the Persians that it cannot be changed, although the world around may change.

The NCRWC has made recommendations for amending the Constitution so as to specifically include in the fundamental rights chapter the judicially deduced rights referred to here.

In countries where fundamental rights are widely violated every day, whether in flouting of labor laws, illegal detentions, discriminatory actions, and other violations, a cynic may well taunt and question the utility of a Bill of Rights. The answer is that it empowers citizens and groups fighting for justice to approach the court and provides opportunities for vindicating the Rule of Law. It also establishes norms and standards that can be used to educate people to know, demand and enforce their basic rights. It has a salutary effect on the administration, which knows that it has to conform to the discipline of fundamental rights. The effort should be to ensure that fundamental rights guaranteed in a Constitution are made living realities for the weak, vulnerable, and exploited sections of society. Moreover, a Bill of Rights is a constant reminder that the powers of the State are not unlimited and that human personality is sacred.

Above all, we should always remember the message of Jefferson that a Bill of Rights is what the people are entitled to against every government on earth, general or particular, and what no just government should refuse.

CHAPTER 8

ANOTHER "SPRINGTIME OF NATIONS"? RIGHTS IN CENTRAL AND EASTERN EUROPE

A.E. Dick Howard

In March 1848, the Lower House of the Hungarian Diet adopted an Address to the Crown. Petitioning the Habsburg king, Ferdinand V, the delegates called for "a national Government, totally independent and free from any foreign influence whatever," a government that "in conformity with constitutional principles, must be a responsible Government, and the result of a majority of the people."[1]

At Pest's Café Pilvax—where law students, intellectuals, and younger politicians gathered—the poet Sandor Petofi and others collected signatures for a petition to be sent to the Diet. At a public meeting, Jozsef Irinyi read the "Twelve Points." Under the title, "What Does the Hungarian Nation Desire?" the petition included demands for freedom of the press, equal civil and religious rights for all, trial by jury, and the taking of an oath to the Constitution by Hungarian soldiers. When the Diet voted final approval of the Address to the Crown, its wording had been strengthened to include at least some of the demands of the Pest petition.[2] Thus, in addition to the demands for national identity and democratic government, the Hungarian reformers sought many of the fundamental principles of a liberal polity.

CENTRAL AND EASTERN EUROPE SINCE 1989

The events of 1989 in Central and Eastern Europe invite comparison to the revolutions of 1848—the "springtime of nations." Both moments inspired great hopes for constitutional government, liberal politics, and individual rights. The movements of 1848, however, were, at least in immediate terms, unsuccessful.

The years since 1989, by contrast, have seen countries in the region, free of Soviet domination, now charting a course whose destination, we hope, embraces constitutionalism, democracy, and the rule of law. Since many of these countries (with important exceptions, such as Czechoslovakia between 1918 and 1938) had little prewar experience with authentic liberal democracy, it becomes especially pertinent to ask what progress has been made in the region since 1989.

In this essay, in speaking of Central and Eastern Europe, I have in mind those European states that were under Soviet rule. I am not including the states formerly part of the Soviet Union (except for the three Baltic states), even though many factors relevant to Central and Eastern Europe would also bear upon developments in Russia, Ukraine, and elsewhere. (I also exclude states like Austria, which, obviously part of "Central Europe" in a historical sense, were not within the Soviet bloc.)[3]

In assessing the prospects for individual rights and constitutional democracy in Central and Eastern Europe, one should realize that the countries of the region share certain characteristics but also differ in important ways. They have in common a half-century of life under communism. All have some legacy of pre-communist European institutions. All are moving, although with varying degrees of commitment and success, toward some form of pluralistic, constitutional democracy. There is a general movement, although again at different paces and results, away from command economics and toward recognition of private enterprise and market economics.

The differences, however, are striking. Various countries have inherited distinct historical and cultural traditions, especially the oft-remarked line separating those lands which were under Habsburg domination before 1918 and those under Ottoman rule. Religious orientations differ—including predominantly Catholic (Poland), Orthodox (Bulgaria, Serbia), or relatively secular (Czech Republic). Some countries are, in ethnic terms, relatively homogenous; others have important national minorities. Some countries are relatively prosperous, even by European standards; others are quite poor. And, as mentioned, the region's countries differ in their pre–World War II experience with democracy.

In the decade and more since 1989, many countries in the region have come a long way in their quest to make liberal constitutional democracy a reality. The accomplishments, at least in the countries which are the region's leaders, include:

— the adoption of new constitutions,
— the advent of separation of powers and checks and balances,
— the holding of free and fair elections,
— progress toward the creation of an independent judiciary,
— the emergence of a flourishing press (although progress in producing independent television has been slower),
— the articulation, especially in the new constitutions, of fundamental human rights, such as expression, religion, privacy, and property,

— a dedication, through treaties, applications for membership in regional organizations (EU, NATO, etc.), and other measures, to the norms of international law,
— economic reform, including the dismantling of command economies and the growth of the private sector.

This list is, of course, only a partial inventory of those factors which might provide benchmarks by which to measure a country's progress toward constitutionalism, democracy, and the rule of law. In a short essay, I cannot undertake a comprehensive review of how well the countries of Central and Eastern Europe are doing in accomplishing their stated goals. I propose instead to consider some of the factors which I find distinctive in appraising the prospects for constitutional democracy and the recognition of individual rights, either in individual countries or in the region in general. Some make the road more difficult; others seem more favorable.

History and Culture

A wag once remarked that Central and Eastern Europe "carries more history in its knapsack than it can consume locally." Certainly the region has been the cockpit of great historical events, especially the ambitions of imperial powers. Poland, for example, in the late eighteenth century, was partitioned among its three powerful neighbors, Prussia, Russia, and Austria, and disappeared from the map of Europe until 1918.

History helps explain the fact that constitutionalism and the rule of law seem more deeply imbedded in some of the post-communist countries than in others. Those lands that lay within the Habsburg domain fell heir to Austrian traditions of a *Rechtstaat*. The Austro-Hungarian Empire may have been notorious for its bureaucratic tendencies, but it developed a strong tradition favoring the rule of law in its successor states in Central Europe. Peoples under Ottoman rule were not so fortunate.

Culture, too, has its place in assessing the region's current prospects. Some observers have argued that countries in which the Orthodox Church is predominant suffer from a close identity between religion, ethnicity, and the definition of the nation, making it more difficult to nurture an open society.

In mulling a country's prospects for liberal constitutional democracy, one should also weigh the importance of that country's s unique historical development. Hungary's traditions of constitutionalism date back at least to its Golden Bull of 1222, a document resembling in many ways England's Magna Carta (1215). Like England, Hungary had (until 1949) the tradition of an unwritten constitution, and even the repressions of the communist era did not erase notions of constitutionalism. It is thus no accident that during the transition to democracy Hungarians placed so much emphasis on principles of legality and that today's Constitutional Court in Hungary sets the pace for the courts of the region.

The Legacy of Communism

From a Western perspective, it is natural to assume that the 1989 revolutions represented a conscious effort to have the people of Poland, Hungary, Czechoslovakia, and other lands not simply join, but rather to *rejoin*, Europe. François Furet has characterized developments in Central and Eastern Europe since 1989 as marking the end of a "long and tragic deviation which had begun in 1917."[4] The notion that the post-communist countries are rejoining their western cousins in a family dedicated to democracy, the rule of law, and an open society is reinforced by the stated goals of the reformers of 1989 and by such documents as the new constitutions in the region. Indeed, Furet's thesis may provide a founding premise for the new democracies.

Half a century of totalitarian rule and of a command economy, however, has inevitably left a deep mark. Most people in the region, of course, did not consent to live under communism. But the habits and mentality inculcated by communist ideology and practices are part of social reality in Central and Eastern Europe. No effort to establish a liberal constitutional democracy can reckon without this residue of the communist era. Among those legacies from the years of communist rule are the weakness of civil society, cynicism, and a distrust of laws and institutions.

The Weakness of Civil Society

When the communists took power, whether in Russia in 1918 or in Central and Eastern Europe after 1945, the new masters set out to destroy those institutions associated with earlier regimes, especially those which might in anyway contest the Party's claim to unquestioned rule. Competing political parties were, of course, banned, and dissenters arrested or executed. But the communists sought more than this; they set out to eradicate the very fabric of non-communist life, like Romans salting the ruins of Carthage. Ideas, values, institutions, loyalties—anything competing with the communist vision had to go.

An important task confronting leaders and people in post-communist Europe, therefore, has been to construct a civil society out of the ruins of the communist era. Václav Havel has been especially astute in seeing the importance of civil society as creating an essential buffer between the individual and the state, a "social space" in which to cultivate a sense of community. It is interesting that efforts to restore civil society to the region have not been without their critics. Václav Klaus fears that proponents of civil society would create new layers of bureaucracy, thus slowing economic reform. This debate suggests a split between those who would identify civil society with NGOs and those who would redefine civil society as identified with a market economy.[5]

Closely allied to the debate over civil society is the effort to create viable local governments in the post-communist countries.[6] Under communism, local units were essentially organs of central administration and certainly not local governments as understood in western discourse. Real power at every level, local or central, was in the Party. As communism collapsed, the creation of local self-government became a central plank in the reformers' platforms, for example, in the demands of Solidarity in Poland. Václav Havel sees an intimate connection between vigorous local government and the prospects for democracy, hoping to see the Czech Republic as a "highly decentralized state with confident local governments."[7] Resistance to devolution, however, is not unique to a communist perspective. National ministries, ever protective of their turf, still prefer centralized power to devolution, despite the arguments that nurturing democracy at the local level heightens a sense of civic participation.

Cynicism and a Distrust of Laws and Institutions

Life under communism bred a sense of loneliness, of distrust of one's neighbors, leading to a Hobbesian disengagement from reliance upon laws and institutions. In the economic realm, shortages of goods and services, made more offensive by favoritism to Party elite, led to the bypassing of institutions and official procedures. Whom you knew, your ability to network, became a part of daily life. In the legal realm, ordinary people came to understand that laws and courts served the Party's purposes. Everyone knew about "telephone justice" the procurator's s picking up the telephone to tell a judge how to decide an important case. "Socialist legality"—law's submission to Party dictates—reigned; one could not speak, save in irony, of a "rule of law."

Skepticism—of government, of parties, of courts, of institutions generally—has carried over into the post-communist era. Even those institutions meant to be representative—political parties, trade unions, parliament itself—are viewed with suspicion. Richard Rose, surveying attitudes in the region, reports that political parties command the trust of only five percent of those polled, and trade unions, nine percent.[8] Likewise, judges and courts must work hard to win the confidence of litigants and others. During the communist era, as in pre-revolutionary France, judges were seen to be spokesmen for the regime, hence discredited and, even after 1989, slow to stir trust and confidence.

The fashioning of new institutions—for example, parliaments and courts—can be accomplished in a relatively short time, as developments in Central and Eastern Europe confirm. But changing popular attitudes, deeply entrenched after decades of communist rule, is a far more daunting task. It is work requiring vigorous civic education and measured, not so much by years, as perhaps by a generation or two.

Elections

Elections do not guarantee the emergence of liberal democracy, but free and fair elections, with widespread participation, are surely one of the prerequisites to any authentic democracy. In this respect, countries in Central and Eastern Europe are, for the most part, beginning to behave like functioning democracies to the West. This pattern was not inevitable; indeed, at the outset, it was movements (such as Poland's Solidarity or the Czech Civic Forum) that emerged with the collapse of communism, rather than western-style political parties. But the movements by and large proved ill-suited to mutate into parties.

A decade after the revolutions of 1989, we see competition between parties—not the same pattern as in Western Europe, but party competition all the same. In most countries in the region, including Poland, Hungary, and the Czech Republic, there has been at least one change of government, sometimes more, through the process of elections. In some cases, former communists have ousted non-communist reform governments. But the one-time communists now compete for power by accepting the constraints of the democratic process, including turning over the reins of power when they, in turn, are defeated. The issues are, of course, not identical to those in other European democracies, but the way in which the political game is played—and the rules observed—suggests that most countries in Central and Eastern Europe are behaving as democracies should.

International and Regional Environment

The presence of aggressive neighbors often dims the prospects for a country's stability. This is especially true in Central and Eastern Europe, where even democratic Czechoslovakia could not survive the predations of Nazi Germany. Today the international and regional climate is far more favorable to the post-1989 democratic transitions. Russia is unstable but hardly the threat it was in Soviet days, and modern Germany is a thriving and healthy democracy, a linchpin of both the European Union (EU) and NATO. The Balkans, of course, tell a far different story, still smouldering in the ashes of the former Yugoslavia.

Aspirations to Join European Institutions

No external force is more of a pull to constitutionalism, democracy, and the rule of law than the aspirations of countries in Central and Eastern Europe to join European and Euro-Atlantic institutions, especially the EU and NATO. Those who seek admission to the EU must show a commitment to democracy, and the EU sees one of its central roles as being the strengthening of democratic institutions.[9] Indeed, in the cases of Spain, Portugal, and Greece, the EU admission was seen as bolstering democracy and guarding against the return of dictatorships in those countries.[10]

Similarly, membership in the Organization on Security and Cooperation in Europe (OSCE) and the Council of Europe reinforce a sense of regional identity, nurturing European norms and values. OSCE's pronouncements, such as the Copenhagen Document (1990), have been influential in the shaping of democratic institutions in post-communist countries.[11] A concrete example of the Council of Europe's work is its Venice Commission, the brainchild of Italy's Antonio La Pergola; the group of experts (professors, judges, etc.) has been active since 1990 in giving advice to individual countries.[12]

Whatever the motivations that impel countries in Central and Eastern Europe to look to Western Europe—economic, political, security, cultural, historical—the forces attracting the newer democracies to emulate standards and norms widely accepted in the more established democracies are powerful. They undoubtedly reinforce the prospects for constitutionalism, democracy, and the rule of law in Central and Eastern Europe.

In May 2004, ten countries were admitted for membership in the EU. Eight are in Central and Eastern Europe—the Czech Republic, Estonia, Hungary, Latvia, Lithuania, Poland, Slovakia, and Slovenia. Other countries, such as Bulgaria and Romania, still hope for admission. Whatever the benefits of EU membership to the countries admitted, it is fair to wonder whether noninclusion will have negative effects on the democratic prospects of those countries left on the outside.

I turn now to the question of rights—how rights are articulated in the constitutions and laws of Central and Eastern Europe and the distinctive way in which rights are viewed in the cultural setting of the region.

NEGATIVE RIGHTS

Since the collapse of communism in Central and Eastern Europe, one of the goals of the new era has been the establishment of liberal constitutional democracy in the countries formerly under Soviet domination. Liberalism requires the establishment of governments that not only are constitutional and democratic, but also recognize certain substantive norms made manifest as constitutional rights.

Ralf Dahrendorf has described three stages in the establishment of liberal constitutional democracy. The first phase involves the drafting and adoption of a constitution, which includes basic rights. Second is the establishment of a market economy. The final stage is bringing about a well-grounded civil society. Dahrendorf thought that the first stage would take six months, the second six years, and the third sixty years.[13] The emerging democracies in Central and Eastern Europe have largely passed through the first stage, although in some countries the process has taken several years (Poland, after extended effort, adopted a new constitution in 1997).

The core constitutional rights that provide the foundation for a liberal democracy may be characterized as being of two kinds, complementary but sometimes in tension: democratic rights and liberal rights. Robert Dahl lists seven political rights associated with the notion of democracy: elected officials, free and fair elections, inclusive suffrage, the right to run for office, freedom of expression, alternative sources of information, and the autonomy of associations.[14]

Rights bolstering liberal values include freedom of speech and association, freedom of conscience, privacy, personal autonomy, minority rights, gender equality, property rights, and freedom of artistic expression. These rights, as Herman Schwartz puts it, are "the basic civil and political rights that were developed in the last two centuries and reflected in such international instruments as the International Covenant on Civil and Political Rights."[15] Some of these rights, notably speech and association, are inextricably linked to the democratic order. Protecting others, however, such as minority rights, often run counter to policies produced by the democratic process. Thus, while democratic rights channel the state's procedural framework, liberal negative rights are meant to place substantive limits on the exercise of political power.

One who consults post-communist constitutions in Central and Eastern Europe finds that their drafters have included, in generous measure, both democratic and liberal rights.[16] Hungary's reformers, in amending the 1949 Constitution, added an impressive list of civil rights and freedoms.[17] Poland's 1997 Constitution includes a catalog of negative rights compatible with, and on occasion more expansive than, international human rights standards. In a complete reversal of the norms of communist constitutional theory, under which a citizen's rights were limited to what the law expressly allowed, Poland's new Constitution makes it clear that citizens may do anything which is not specifically prohibited by law. Irena Grudzinka Gross believes this to be "the Constitution's most important norm. For the first time in Poland's history, the boundaries between state power and the citizen's civil liberties are clearly drawn."[18]

The status of rights in the Czech Republic presents an intriguing ambiguity. In 1991, before the split between the Czech Republic and Slovakia, Czechoslovakia adopted the Charter of Fundamental Rights and Freedom, a document written with considerable American and Western European input and containing an extensive set of rights and freedoms. Since the "velvet" divorce, each republic has its own constitution. While the Slovak document expressly includes the Charter in the body of the Constitution, the Czech Constitution is more oblique; it makes the Charter "a part of the constitutional order of the Czech Republic."[19] Some observers have surmised that this imprecise formula reflects a view, held by some Czechs, that the Charter was not altogether a voluntary enactment of the Czech people but represented pressures from the West, especially from Brussels and Strasbourg.[20] Another theory is that free-market proponents among the Czech Constitution's framers were concerned about the lingering views of former communists who wanted social and economic guarantees among the rights having constitutional status.[21]

The debate over the status of rights in the Czech Constitution suggests a larger question about rights in the region's constitutions generally. Undoubtedly, the new constitutions catalog, more or less consistently, the core negative rights considered indispensable to liberal democratic government. But one may well ponder to what extent these provisions represent the considered, unrestrained judgment of the region's several countries. Most of these countries are anxious to gain membership in the EU. The EU's accession requirements include the Copenhagen political and economic provisos and the *acquis communautaire* (80,000 pages detailing the laws, norms, and standards enforced in the EU countries).[22] One can only surmise just what paths constitutional framers in individual Central and Eastern European countries might have taken but for the pressure to satisfy Western European expectations. Alina Mungiu Pippidi sees an "affront to common sense" in that "countries as different as Lithuania, Romania, and Hungary have relinquished their elementary right of choice, or at least their right to tailor their choices to their own special needs."[23] Ultimately, of course, the test of whether truly liberal constitutional democracy becomes a reality in post-communist Europe depends in large measure upon the extent to which the rights spelled out in the constitutions become internalized and enforced.[24]

An evaluation of post-communist constitutions invites the question whether rights are conceived as natural rights, preceding political society, or as rights grounded in positivist theory, deriving from constitutions and laws. Socialist constitutions, following the Soviet model, typically had extensive catalogs of rights. But the constitution was not seen as imposing an extra-political constraint on state policies. Rights were not a barrier to state power but, in effect, benefits left to the state's discretion. The exercise of rights depended upon the fulfillment of the citizen's duties. Rights could not be exercised to the detriment of state interests. Rights thus were not self-executing; the state decided when and how those norms would be effective.[25]

Many of the post-communist constitutions, by contrast, seem to revive the idea of rights as representing objective values, that is, being pre-political natural rights. Hungary's Constitution recognizes "inviolable and inalienable fundamental human rights."[26] Gabor Halamai sees this provision as rejecting the socialist notion of rights as "privileges granted due to particular considerations by state power."[27]

It is not clear, however, whether a natural law approach in fact underlies the region's constitutions in general. Romania's Constitution simply states that "citizens enjoy the rights and freedoms granted to them by the Constitution and other laws...." Is this positivism? One reason why it is difficult to determine the extent to which drafters were thinking in natural law or positivistic terms is that the exercise of drafting the new bills of rights was, in may ways, a synthetic exercise. That is, drafters depended heavily upon existing western constitutions and also on regional or international documents as sources on which to draw in shaping a bill of rights. It is not my impression that drafters necessarily did the intellectual heavy lifting implicit in a debate over the nature and origin of rights, as opposed to deciding which rights, and in what language, ought to be included.

Another question raised by examining the text of post-communist constitutions is the extent to which negative rights are limited or qualified. Like the German Basic Law, Poland's Constitution allows for limitations to be placed on basic rights but states that such limits may only be achieved by statute (thus, not by executive orders or other substatutory enactments) and only in accordance with Article 31.3, which provides

> any limitation upon the exercise of constitutional freedoms and rights may be imposed only by statute, and only when necessary in a democratic state for the protection of its security or public order, or to protect the natural environment, health or public morals, or the freedoms and rights of other persons. Such limitations shall not violate the essence of the freedoms and the rights.[28]

In Hungary, the Constitutional Court has drawn a distinction between "core" and "peripheral" rights.[29] Where a basic right is involved, a restriction, to be upheld, must be justified by a compelling reasons, for example, the enforcement of another basic right. The legislature must choose a course of action which is least restrictive of the right, and the enactment will be reviewed under standards of proportionality.[30]

One of the most interesting and important tests of how rights are understood in Central and Eastern Europe is the application of libel and slander laws to limit freedom of expression. Especially troubling is the use of such laws to punish individuals for criticizing high government officials, such as the president. An article of the Polish penal code criminalizes "publicly insulting the Polish Nation or State or its system of supreme bodies." Mark Brzezinski has lamented the willingness of Polish prosecutors to use defamation laws in ways that undermine freedom of expression.[31] For example, a journalist was imprisoned under Article 270 for publishing an article referring to the local voivodship council and to Solidarity trade union officials as "dopes" and "small time politicos and careerists."[32]

Hungary's Constitutional Court, an activist court by any standard, has held that the standard that public officials must meet when complaining of libel is higher than that which confronts ordinary citizens.[33] Moreover, the Court said that opinions reflecting value judgments must be protected even if they hurt the honor or reputation of government officials.[34] Such decisions are closer to the standard of "actual malice" laid down by the United States Supreme Court in *New York Times v. Sullivan*[35] and show an awareness of the vital role played by robust debate in a free society. In general, however, the way in which defamation laws are used in Central and Eastern Europe are inconsistent both with the spirit of liberal democracy and with regional and international norms, such as the European Convention of Human Rights.[36]

POSITIVE RIGHTS

The universe of rights is not limited to those rights ("negative" rights) that put space between government and the individual. In the making of modern constitutions, it has become increasingly common for drafters to include positive or affirmative rights—rights which may be read as placing a mandate on government to promote the citizens's welfare. Examples include the right to work, the right to an education, and the right to social security or welfare benefits. Such rights are sometime called "second generation" rights.[37]

Early instances of the inclusion of social and economic rights in a constitution include Mexico's 1917 Constitution and the 1919 Constitution of the Weimar Republic. The revolutionaries who drew up the Mexican Constitution enumerated in great detail a program of labor and social welfare.[38] The Weimar Constitution was the first European constitution to include second-generation rights; it did so in rather more general terms than did the Mexican document.[39]

Constitutions drafted in post-communist Central and Eastern Europe present an unusual context in which positive rights have been articulated. In western liberal democracies, negative rights typically were well established before the emergence of second-generation rights. By contrast, in countries within the Soviet sphere, positive rights were, at least in formal terms, well established; now those countries are attempting the adoption and enforcement of negative rights alongside the positive regime associated with socialist theory. Communist constitutions were not, however, a meaningful limitation on state power; certainly, citizens could not look to independent courts to enforce either positive or negative rights.

The post-communist constitutions of Central and Eastern Europe include extensive provisions for social and economic rights. At least two factors account for this aspect of constitutionalism in these countries. One is the legacy of the socialist period, when the prevailing dogma expected individuals to look to the state for their economic wherewithal. The other derives from the traditions of social democracy in Europe, where the role of the state has typically been more extensive than is true in the American polity.

The rights of workers furnish a good example of the provision for positive rights in the constitutions of Central and Eastern Europe. Hungary's Constitution guarantees workers the right "to rest and leisure and to regular holidays, with pay."[40] The 1997 Polish Constitution provides for the right to choose and pursue the occupation of one's choice, a minimum compensation, safe and hygienic working conditions, and vacations and paid holidays.[41] The Czech and Slovak constitutions offer an interesting contrast with each other. In the Czech Republic, where sometime prime minister Vaclav Klaus was often thought of as Central Europe's Margaret Thatcher, the Czech Charter sets out a general principle of fair remuneration for work and satisfactory working conditions.[42] In Slovakia, a more conservative country, where the socialist legacy runs deeper, the Constitution spells out much more explicit and extensive rights for workers, including "fair and satisfactory conditions at work," as well as the

> right to compensation for the work performed to secure a dignified standard of life, the security from arbitrary dismissal and discrimination at work, the setting of maximum working hours, reasonable rest time after work, a minimum amount of paid vacation, the right of collective bargaining.[43]

Another cluster of positive rights in the Central and European constitutions consists of those offering social security and welfare. For example, the Hungarian Constitution provides for a right to social security and entitles citizens to subsistence in the event of illness, disability, widowhood, old age, or unemployment.[44] Education is typically enumerated among the constitutional rights in the region.[45] A modern note is struck by constitutional provisions recognizing "third generation" environmental rights. The Czech Charter declares that "everybody has the right to live in a favorable living environment and is entitled to timely and complete information about the state of the living environment and natural resources."[46]

How does one enforce a positive right? Negative rights are typically assumed to be self-executing, that is, that they do not require legislative implementation before a court can fashion an appropriate remedy, such as an injunction. Enforcing a negative right commonly entails telling the government what it may *not* do. But the enforcement of positive rights is more complicated. Such rights often require an assessment of empirical data, professional standards, or policy judgments better left to legislatures or other bodies.

The problem of enforcement becomes especially challenging when a claim of right bears upon the allocation of resources. Resources are not infinite. Legislators often must choose among worthy purposes. Spending for one purpose, such as education, may leave less money to spend on other purposes, such as welfare. Some would argue that deciding among competing ends for which money may be spent is a quintessential legislative chore. Moreover, judges who undertake to implement positive rights may find themselves acting, in effect, as administrators, supervising government functions, perhaps for years (as has been the case with much institutional litigation in American courts).

Constitutional drafters have sometimes tackled this problem by dividing rights into two categories, those intended to be justiciable, and those which are not. Those who fashioned Ireland's 1937 Constitution used "Directive Principles of State Policy" to delineate social and economic goals.[47] Article 45 of that constitution set out to make these goals nonjusticiable:

> The principles of social policy set forth in this Article are intended for the general guidance of the Dail [parliament]. The application of these principles in the making of laws shall be the care of the Dail exclusively, and shall not be cognizable by any Court. . . .[48]

The drafters of India's 1950 Constitution followed the Irish example when they precluded the courts from reviewing the Directive Principles of State Policy. These principles, drawn in broad terms, embody the drafters' ideas of India's legislative goals, including the creation of a welfare state, the creation of a classless and casteless society, and India's promotion of international peace.[49]

Constitutions in the communist world were little more than facades, and it is not surprising, therefore, that drafters of post-communist constitutions have had little direct experience with questions of the justiciablility of constitutional norms. Peter Paczolay has observed that Hungary's drafters "were not aware of the importance of having a precise and theoretically well-founded regulation explaining the scope of social and economic rights, and the participants in the political negotiation also neglected this question."[50]

The region's constitutions seem to differ in their approaches to the question whether positive rights are meant to be self-executing or require legislative implementation. Social, economic, and cultural rights in the Czech and Slovak republics are seen as merely aspirational.[51] As one pair of commentators put it, the enforceability of these rights "can be based only on laws that transform these 'soft-law' guidelines into 'real' rights."[52]

Poland's 1997 Constitution seems to distinguish between those rights which are enforceable and those which are not. Thus there is the categorical statement, "Everyone has the right to safe and hygienic conditions of work," whereas the goal of full employment reads as a policy statement: "Public authorities pursue policies aimed at full, productive employment by implementing programs to combat unemployment."[53]

Where a constitution makes no evident distinction between enforceable and nonjusticiable rights, it is left to the court to work out the problem. The Hungarian Constitution, in Article 70/E, contains social and economic rights framed in terms implying enforceability:

(1) The citizens of the Republic of Hungary shall have the right to social security; they shall be entitled to provision necessary for subsistence in case of old age, illness, disability, widowhood, orphanhood, and unemployment owing to circumstances beyond their control.
(2) The Republic of Hungary shall realize the right to this provision through social insurance and social services.[54]

Peter Paczolay objects to the seemingly categorical nature of this language; it "misleadingly suggests that the right to welfare services should be of the same character as basic political rights and liberties—that is, enforceable by judicial means." In his judgment, the drafters of this provision failed to distinguish between those rights which are innate and inalienable and those which are "tasks of the government to fulfill to the degree reasonably permitted by economic conditions."[55]

The question of the enforceability of social and economic rights surfaced in Hungary's Constitutional Court as early as 1990. In a concurring opinion, Chief Justice Solyom denied that the right to social security is self-executing. Article 70/E, he said,

does not entitle anyone to have a right to "social security and safety"; legal claims on such a general level cannot be defined. . . .[I]t may be stated that social rights are not inherent rights but state tasks. Neither the extent, nor the criterion, of social care are specified by the constitution; this and the related practical implementation are the responsibility and duty of the legislature and of the executive.[56]

Despite such reasoning, constitutional courts in Central and Eastern Europe do find occasions to enforce social and economic rights. Indeed,

Hungary's Constitutional Court in 1995 struck down 26 provisions of the government's austerity package. The plan, found to conflict with constitutional guarantees, would have restructured and cut sick-leave benefits and made substantial changes in long-term maternity and child-care benefits.[57]

Poland's Constitutional Tribunal has relied on principles of "social justice" and constitutional guarantees to enforce social and economic rights. For example, in 1993, the Tribunal struck down a provision of an act which would have dramatically limited the periods for which benefits would be paid to unemployed workers. The Tribunal found the act as violating the rules of just distribution of goods necessary to meet elementary human needs, hence inconsistent with a constitutional principle of social justice. It also infringed the state's duty to protect motherhood and the family, as well as state duties regarding unemployed persons ready and willing to work, implicit in the constitutional guarantee of a right to work.[58]

Poland and Hungary offer the best tests of judicial willingness to enforce social and economic rights. There are perhaps fewer cases than one might have expected in other countries in the region. It may be that austerity programs of the kind that sparked litigation in Hungary and Poland have not been pushed in other countries. Or it may be that citizens have not yet got used to going to court to protest rights violations, especially in the economic sphere.

Even if one concludes that social and economic rights ought not to be justiciable, it does not follow that a constitution should be silent on such rights. Allowing legislators to give content to such goals may make practical or jurisprudential sense. But, given the ethical and social insights of modern life, there is force in the argument that, without respect for basic social and economic justice, others rights may seem empty and insufficient.

NATIONAL MINORITIES

In the nineteenth century, nationalism was commonly seen as a liberal force, as in the revolutions of 1848. But much of the history of the nineteenth and twentieth centuries reveals the darker side of nationalism. Coupled with language, religion, culture, or other ways of distinguishing "us" from "them," nationalism can become an awesomely destructive force, subversive of human rights (especially those of national minorities) and of liberal institutions generally.

Nationalism—often inflamed by opportunistic political leaders such as Serbia's Miloševic, Croatia's Tudjman, and Slovakia's Meciar—played a key role in destroying the two multinational states created by the Trianon settlement, Czechoslovakia and Yugoslavia. In the unitary states of Central and Eastern Europe, the presence of significant national minorities poses a special challenge to democratic stability. If ethnicity becomes the basis for political parties, national minorities may well become permanent political minorities.

Faced with the threat that their nations might disappear altogether as the result of assimilation policies implemented by the imperial powers (Prussia, Russia, and Austria), nationalists in nineteenth-century Central and Eastern

Europe embraced the distinction between "mere states" and "political nations." The idea was that the cultural identity of a nation could live on despite the loss of statehood. Whatever its romantic appeal, this aspiration presents several problems. One is that some ethnic groups (e.g., the Roma) have never been represented by a historical state. Another is that the populations of the historical nations are scattered throughout the region, making the drawing of rational political borders often quite difficult. It was just such a problem that contributed to the undermining of Woodrow Wilson's policy of self-determination after World War I. In 1848, the Romanian patriot Balcescu declared, "Until a people can exist as a nation, it cannot make use of human liberty."[59] This is lofty language, but it raises the concern that an ethically diverse people cannot exist as a nation without forced assimilation.

Approaches to the problem of the rights of national minorities vary among the countries of Central and Eastern Europe. One factor is the presence or absence of large ethnic minorities; homogeneous states tend, by and large, to have more liberal policies, while countries having large national minorities are more apt to see claims made by these minorities as threatening secession. Moreover, those who frame constitutional or legal provisions for the rights of national minorities must decide whether rights are individual or collective. OSCE's influential Copenhagen document (1990) devotes perhaps a quarter of its length to provisions regarding the rights of national minorities, but the document is unclear as to whether the rights are those of individuals, acting collectively, or collective rights of communities.[60]

Of the countries in the region, Poland and Hungary appear to have the most extensive guarantees for minorities. It is probably no coincidence that they are also the most ethnically homogeneous states in the region. Poland's 1997 Constitution provides:[61]

1. The Republic of Poland ensures to Polish citizens belonging to national or ethnic minorities the freedom to maintain and develop their own language, to maintain customs and traditions and to develop their own culture.
2. National and ethnic minorities have the right to establish educational and cultural institutions, institutions designed to protect religious identity, as well as to participate in the solution of matters connected with their cultural identity.

The Hungarian Constitution provides for both the special and equal treatment of minorities. The provision for special treatment is framed in explicitly collective terms:[62]

Sec. 2: The Republic of Hungary shall protect national and ethnic minorities. It shall ensure their collective participation in public life, foster their culture, the use of and the instruction in their native languages, and the right to use their name in their own language.

Sec. 3: The Acts of the Republic of Hungary shall ensure the representation of the national and ethnic minorities living the territory of the country.

Sec. 4: The national and ethnic minorities may establish local and national self-governments.

Equal treatment is guaranteed through an anti-discrimination provision:[63]

Sec. 1: The Republic of Hungary shall ensure human and civil rights for everyone within its territory without discrimination of any kind, such as upon race, color, sex, language, religion, political or other opinion, national or social origin, property, birth or upon any other grounds.

Sec. 2: Any discrimination of people as described in section (1) shall be severely punished by law.

Indeed, the Hungarian Constitution goes so far as to authorize affirmative action:[64]

The Republic of Hungary shall promote the attainment of the equality before the law also by measures aimed at eliminating inequalities of opportunity.

Building upon the foundation provided by the Constitution, the Hungarian Parliament, in 1993, adopted the comprehensive Act on the Rights of National and Ethnic Minorities.[65] The statute identifies both individual and collective rights. Among the former are the rights to equal chances (including positive discrimination), the right to establish ethnic associations, parties, and organizations, and the right to observe minority traditions. More striking is the act's enumeration of collective rights: the right of minorities to preserve, cultivate, and transmit their identity as a minority, to cultivate their traditions and language, to establish self-governments at the national and local levels, to develop a national network of education, training, cultural, and scientific institutions, to establish international relations, and to be represented in parliament.[66] Just how to implement a right of collective representation in parliament has still not been worked out; Hungary's parliament has a single chamber whose members are elected by single constituency and party lists, leaving no apparent place for corporate solutions.[67]

Local self-government appears to be the main focus of the Act's approach to achieving the rights of national minorities. Within two years of the Act's adoption, more than 800 minority self-governments had been established in Hungary.[68] The greatest obstacle to genuine local self-government for minorities (indeed, as it is to local self-government in general in Central and Eastern Europe) is the lack of adequate financing.[69]

Hungary's laws for advancing the interests of national minorities are admirably enlightened. In enacting these laws, however, Hungary is looking beyond its borders, hoping to set standards for protecting minority rights, which will be applied to its own kinsmen who live in Romania, Slovakia, and Serbia (about 10,000,000 Hungarians live in Hungary itself; another 5,000,000 live in neighboring countries). Indeed, Hungary's overriding concern with ethnic Hungarians elsewhere is confirmed by the enactment in

2001 of a minority "status law." Effective on January 1, 2002, the law provides rights and preferences for ethnic Hungarians living outside of Hungary, including the right to work in Hungary for three months each year.[70] The law aims at preserving the cultural identity of Magyars wherever they may live. Romania, complaining bitterly about the law, challenged it before the Council of Europe. The Council upheld the act but urged Hungary's government to address other governments' concerns.[71]

Because Hungary is ethnically relatively homogeneous, its laws are not the best test of the status of national minorities in the region generally. More diverse countries pose a greater challenge, and in those places progress has been more mixed.

Even the relatively liberal and progressive Czech Republic offers an example of the problems of minorities. After the "velvet divorce" of January 1, 1993, thousands of Slovaks applied for Czech citizenship. The Czech Parliament responded by enacting a law allowing Slovaks to become Czech citizens only if they had resided in the Czech Republic for two years, had committed no felonies during the previous five years, and had renounced their Slovak citizenship. The law also required fluency in the Czech language. Because of the felony provision and the close similarity between the Czech and Slovak languages, it seemed that the law was a thinly veiled attempt to preclude the Roma from gaining citizenship. Following international criticism, the law was subsequently modified.[72]

Slovakia's Constitution has guarantees for national and ethnic minorities similar to that of the Czech Republic's charter.[73] But another provision of the Slovak Constitution stipulates that the "exercise of rights by citizens of a national minority guaranteed by this Constitution may not threaten the sovereignty and territorial integrity of the Slovak Republic or discriminate against other citizens."[74] This language obviously is meant to prohibit separatist movements and also appears to forbid affirmative action.[75] The nationalist flavor of the Slovak Constitution is reflected in the language in the Preamble that states: "Slovak is the state language in the territory of the Slovak Republic. The use of other languages in dealings with the authorities will be regulated by law."[76]

In 1995, responding to international pressure, Hungary and Slovakia signed a bilateral treaty by which Hungary recognized the current border between the two countries and the Slovak Republic promised linguistic, educational, and representation privileges to its Magyar minority. The Slovak National Council ratified the treaty, but it simultaneously adopted two declarations, one rejecting the idea of collective rights for minorities, the other opposing any form of local self-government based on ethnicity.[77] Magyar complaints have focused on such issues as the need for bilingual road signs in areas with a majority Hungarian population, the right to use Hungarian names, and a wish to have autonomy in education. Conditions have improved as a result of legislation adopted by the Slovak Council aimed at gaining admission to the Council of Europe.[78]

Bulgaria illustrates how sensitive a majority people can be about what they view as the "fifth column" potential of national minorities. Bulgaria's Constitution contains guarantees for individual rights and for equality but eschews collective rights, indeed, rejecting the term "minority" altogether.[79] Moreover, the Constitution, while providing for the rights to peaceful and unarmed assembly and the right to form associations, restricts those rights in ways obviously aimed at ethnic minorities, especially the Turks. The Constitution bans organizations which threaten the country's territorial integrity or unity or which incite racial, ethnic, or religious hatred.[80] The Constitution explicitly bans the establishment of ethnically based political parties and forbids citizens' associations, including trade unions, from engaging in political activities.[81] Such provisions, especially in the hands of unfriendly government officials, obviously impact the ability of Turks and other minorities to have a meaningful role in Bulgarian politics.[82]

Minorities find it helpful to have a "mother country," which will use international forums and other vehicles for protecting their interests. Magyars in Slovakia and Romania have obviously been the beneficiary of Hungary's activities on their behalf. Ethic minorities, such as the Roma, who lack such patrons have a harder time of it. Roma suffer not only for a bad image in the eyes of other peoples, they also are usually poor and badly educated. Polls reveal the negative views others have of the Roma.[83] Roma frequently complain of rank discrimination in education, employment, housing, banking, and the system of justice.[84] Roma families in the Czech Republic, complaining that their children were shunted off into special schools for the mentally disabled and socially maladjusted, sought relief in the Constitutional Court. When that tribunal rejected the complaint, the plaintiffs took the case to the European Court of Human Rights in Strasbourg.[85]

Looking at Central and Eastern Europe generally, it is paradoxical that in many of the region's countries, the national question is less troubling than it was before World War II because of the traumatic upheavals, displacement of populations, and boundary changes which came in the wake of the war. For example, where minorities comprised one-third of the population of prewar Poland, that country is now largely homogeneous. Germans and Jews—and now even Slovaks—dwell in insignificant numbers in today's Czech Republic. And a key reason why Slovenia was able to extract itself from a collapsing Yugoslavia was the fact that so few Serbs live there. It would be untenable (not to mention offensive) political science to suppose that democracy can exist only where there is ethnic homogeneity. Yet in Central and Eastern Europe, it appears that, where minority populations are sufficiently small to allay historic fears of irredentism, that fact makes it easier for such countries to spend more time focusing on other problems, such as the economy.

Constitutional Courts

As constitutionalism takes shape in the post-communist countries, no phenomenon is more striking than the creation and emergence of constitutional courts. These courts, having (among other competences) the power to strike down acts of parliament as unconstitutional, represent two important departures from the role of courts in Central and Eastern European countries before 1989.

First, the very idea of a constitutional court—typically modeled on the German Constitutional Court—emphatically rejects a basic premise of socialist constitutionalism. In the communist era, there was a unity of state power, and the Party called the shots. Authentic judicial review, like the separation of powers, represents an idea that is anathema to socialist ideology.

Moreover, the emergence of judicial review moves away from European traditions of parliamentary supremacy and the judge as civil servant, dedicated to seeing that parliament's will is carried out. Hans Kelsen provided for a constitutional court in Austria's 1920 Constitution, but the idea saw little imitation until after World War II, when the creation of the German Constitutional Court provided a powerful model for other countries. Since the collapse of communism, Central and Eastern European countries, with various degrees of boldness, have embraced the idea of a constitutional court.

In 1986, three years before the Roundtable talks, Poland's communist government took the region's first steps towards the creation of constitutional courts. That step was modest but important. The Constitutional Tribunal was given the power to rule on the constitutionality of government actions, but, while it could rule on acts of Parliament, its rulings were not final. Leszek Garlicki observed that the Tribunal was "designed not to supervise Parliament, but to help Parliament maintain its position as the country's supreme legislator."[86]

After 1989, Poland moved to strengthen the powers of the Tribunal. Amendments to the enabling statute, adopted in 1989 and in 1992, broadened the Tribunal's jurisdiction. The 1997 Constitution went even further. Before 1997, individual citizens did not have standing to bring constitutional challenges before the Tribunal. The new Constitution explicitly provided for constitutional complaints:

> Everyone whose constitutional freedoms or rights have been infringed shall have the right to appeal to the Constitutional Tribunal for its judgment on the conformity to the Constitution of a statute or another normative act upon the basis of which a court or organ of public administration has made a final decision concerning his freedom or rights or his obligations specified in the Constitution.[87]

Even more critical to the Tribunal's function was the elimination of the Sejm's power to overrule a decision of the Tribunal.[88] Thus Poland has made the Constitutional Tribunal not simply the guardian of Parliament, but the guardian of the Constitution.

Hungary's Constitutional Court has gained a reputation as the region's most activist tribunal. That court has proved unafraid of risking confrontations with parliament; in the Court's first five years, about one-third of all laws challenged before the Court were found to have at least one unconstitutional provision. In its first ten years, the Court struck down more than 100 statutes—an activist pace which exceeds even the Warren Court in its most exuberant moments.[89]

The Hungarian Court has also been willing to take stands clearly counter to public opinion. For example, the Court declared the death penalty to be unconstitutional, even though, as László Sólyom has commented, "the vast majority of the Hungarian population views this punishment as acceptable and necessary."[90] Similarly, despite considerable public outcry, the Court invalidated a law permitting the government to prosecute certain crimes that—for political reasons—had not earlier been prosecuted and were technically barred by the statute of limitations.[91]

The region's constitutional courts are not always models of courage. In the summer of 1998, Slovakia's parliament banned the coverage of elections by private media, leaving coverage to media dominated by Vladimír Mečiar and giving his party a deciding advantage in the pending election. Opposition deputies promptly challenged the law, but a few weeks before the election the Court's chairman said that the matter needed extended consideration and therefore the court would not deal with the case before the election. Six months later, well after the election, the tribunal finally decided that the law was unconstitutional.[92]

Post-communist constitutional courts vary in their jurisdiction, for example, in whether they can entertain citizen's petitions. They vary in their willingness to stand up to political forces and be a serious check on legislative and executive power. However, there is little doubt that, in general, the new constitutional courts are a signal force in nurturing constitutionalism and the rule of law. László Sólyom, formerly Chief Justice of Hungary's Constitutional Court, declares that the Court has "made both the politicians and the population conscious of the secure protection of constitutional rights" and that politics must operate "within the framework of the Constitution—not vice versa, as before, when the law was conceived as merely a political tool."[93]

Of the constitutional principles enforced by Hungary's Constitutional Court, none is more central than the rule of law. As shaped by the Court, pillars of the law include legal certainty—the requirement that laws be clear and unambiguous—and the principled coherence of the Constitution.[94] Hungary can claim, with some justification, that its revolution has been a revolution under the rule of law.

A CULTURE OF RIGHTS

Thomas Masaryk understood that, if democracy is to survive and flourish, it requires "the constant development and extension of public spirit." Masaryk saw individualism as the handmaiden of democracy. However, he added, that does not mean a "capricious individualism," but rather a sense of individual responsibility.[95]

I commented, earlier in this essay, on the destruction of civil society during the communist era in Central and Eastern Europe. Rebuilding civil society is one of the prerequisites to promoting liberalism and the rule of law. Ultimately, whether constitutional liberal democracy will take root in the region depends upon there being public faith in the justness of liberal rights as against the state, and a belief that the state's institutions can be influenced in ways that will advance those rights.[96]

Cynicism regarding political institutions and the rule of law is one of the greatest obstacles to the development of a constitutional and legal culture of rights. As Herman Schwartz has observed, building faith in law and political institutions is especially difficult because of a pervasive "legal nihilism, the conviction that legal rules have almost no influence on how officials behave." Ewa Letowska recounts an anecdote illustrating cynicism regarding constitutional rights:

> A hungry traveler walks into a shady restaurant in Moscow. He sits down and inspects the menu. "I'll have the pork chops," he says. "We don't have any," answers the waiter. "Well then, I'll have the meat balls." "We don't have those either." "How about liver then?" "Nope," answers the waiter. The annoyed customer finally asks: "Am I reading the menu or our constitution?"[97]

Communism's legacy of cynicism is extended by the way in which the old nomenklatura were often able to use their old positions and connections to attain preferred opportunities in the new economy. Further, it does not help that so many people in Central and Eastern Europe still rely heavily on the welfare state, breeding heightened feelings of insecurity as the command economy gives way to a more market-oriented system. Apprehensive and uneasy, citizens look for scapegoats—criminals, the idle, the unemployed, gypsies, Jews, foreigners. There is the danger that the culprits, thus identified, may be seen as not deserving of the full range of human rights.[98]

When Americans wrote their state and federal constitutions, they had the advantage of having, in effect, already practiced the uses of democracy and constitutionalism. Anyone who reads the tracts and resolutions of the 1760s and 1770s can see a constitutional mentality already at work.[99] It is possible to argue that, at the American founding, the culture of constitutional democracy preceded the making of the Constitution.

In Central and Eastern Europe, as Irena Grudzinska Gross has observed, the cart of constitutional democracy has been placed before the horse of civil society and a legal culture: "These parties and the state were left to build civil society, which elsewhere had been the foundation, not the product, of democracy."[100]

One threat to the people's internalization of a culture of rights is the perception that norms are being imposed from without the country. Before 1989, the people of Central and Eastern Europe had not yet domesticated western ideas of human rights, but they were well acquainted with nationalism as a means to political cohesion. Thus opportunistic politicians can impugn notions of "universal" rights as alien and as a threat to national identity. As Andras Sajo has argued:

> [N]ationalism is the most common and most powerful anti-legalistic ideology. Irrespective of the nature of nationalism (which ranges from a desire to protect one's nation's interests against multinational imperialism to extreme racism), one has to understand that nationalism is a constant source of hostility to human rights and even to the rule of law. Throughout the region, both rights and legality are now presented as imposed values. Some of the elements of the human rights/rule of law "package" are singled out for special obloquy, because the rights in question are thought to impugn national virtues.[101]

Another barrier to the development of a liberal legal culture is seeing rights as being simply means to an end, such as material well being, rather than worthy of respect as such. Despite all the public debate over the drafting of a new Constitution for Poland, when the referendum finally took place, in 1997, only 43 percent of the voters turned out. Ewa Letowska surmises:

> In reality, the referendum lacked popularity, because many citizens realized that a new constitution would not spell swift and direct improvement in their standard of living. Only a few years ago, as was recalled, the public's belief in the constitution's direct importance in their lives had been much stronger. But with time this conviction dissipated. While support waned, paradoxically, economic conditions improved. By the beginning of the year, the majority had concluded that one can live relatively decently without a new constitution. And the ugly pre-referendum campaign fed the conviction that the Constitution was essentially for "politicians," not for the "people."[102]

Andras Sajo summarizes the challenges to the development of a culture of rights and a strong civil society in Central and Eastern Europe:

> Neither Eastern European nor Western societies are ready to undertake the self-restraint, self-help, self-organization, and sacrifice envisioned by the civil society mythology. Nor are they to be judged or blamed for this lack of readiness. Not only are the material preconditions of individual independence missing in Eastern Europe, but so are trust and belief in common causes. It is hard to expect charity from first-generation murderers and tax evaders who were formed in the illusionless world of communism and were then tossed willy-nilly into the postcommunist free-for-all. The strategic invocation of "civil society," considered to be what Western Foundations want to hear, leads to the financing of movements and groups which either become parasites on the donors or who cynically take the money and run.[103]

None of this is to say that a legal and constitutional culture will not take shape in Central and Eastern Europe, only that this development, devoutly to be hoped for, must travel a rougher road than many in the West might realize. There are encouraging signs, for example, the respect Hungarians have for the Constitutional Court; opinion polls routinely report the tribunal to be the most popular organ of government.[104]

Ultimately, some form of civic education is required to nurture healthy liberal constitutional democracies. A people who do not understand and act upon the precepts of free government and fundamental rights are unlikely to keep those rights alive and well. Thomas Jefferson understood the connection. In his *Notes on the State of Virginia*, Jefferson described his Bill for the More General Diffusion of Knowledge (1777). Its purpose, he said, was "rendering the people the safe, as they are the ultimate, guardians of their own liberty.... Every government degenerates when trusted to the rulers of the people alone. The people themselves therefore are its only safe depositories."[105]

Those people must, of course, learn and practice the traits which accompany a regime of rights, of liberalism, of constitutional democracy. Virginia's 1776 Declaration of Rights declares: "That no free government, or the blessing of liberty, can be preserved to any people but by a firm adherence to justice, moderation, temperance, frugality, and virtue, and by a frequent recurrence to fundamental principles."[106] This language, written for Americans in the eighteenth century, is a fair signpost for Central and Eastern Europe in the new era.

NOTES

I am grateful to John Anderson for his seasoned research assistance in the preparation of this essay.

1. Great Britain, House of Commons, Seasonal Papers, 1851, vol. LVIII, "Correspondence: Relative to the Affairs on Hungary, 1847–1849," 42–44, quoted in Edsel Walter Stroup, *Hungary in Early 1848: The Constitutional Struggle against Absolutism in Contemporary Eyes* (Buffalo: Hungarian Culture Foundation, 1977), 65–69.

2. Stroup, *Hungary in Early 1848* (Buffalo: Hungarian Culture Foundation, 1977), 100, 102.

3. See Timothy Garton Ash, "The Puzzle of Central Europe," *New York Review of Books* (New York: A.W. Ellsworth, March 18, 1999) (discussing the difficulties and implications of one's definition of Central Europe).

4. See Ralph Darendorph, *Reflections on the Revolution in Europe* (London: Chatto & Windus, 1990) quoted in Harald Baldersheim and Michael Illner, "Local Government: The Challenges of Institution-Building," in Harald Baldersheim et al., eds., *Local Democracy and the Processes of Transformation in East-Central Europe* (Boulder, CO.: Westview Press, 1996), 6.

5. See Jiri Pehe, "Civil Society at Issue in the Czech Republic," *RFE/RL Research Report* (Munich, Germany; New York, N.Y.: Radio Free Europe/Radio Liberty, Inc., 1994), 13.

6. See A.E. Dick Howard, "After Communism: Devolution in Central and Eastern Europe," 40 *South Texas Law Review* (Houston, TX.: South Texas College of Law, 1999), 661.

7. Václav Havel, "From the Czech Republic with Love and Dreams: Václav Havel's Vision of a Post-Communist Society," *Utme Reader* (Minneapolis, MN: Lens Pub. Co., May–June, 1993), 90.

8. Richard Rose, "Another Great Transformation," *Journal of Democracy* (Washington, D.C.: National Endowment for Democracy, January, 1999), 54.

9. The European Council's 1993 Copenhagen Council declared that member-ship in the European Union "requires that the candidate country has achieved stability of institutions guaranteeing democracy, the rule of law, human rights, and respect for the protection of minorities. ..." Copenhagen European Council, June 21–22, 1993, Conclusions of the Presidency, Bulletin of the European Communities, 13.

10. See Martin Holland, *European Integration: From Community to Union* (London: Printer Publishers, 1993), 167.

11. See Conference for Security and Cooperation in Europe, *Document of the Copenhagen Meeting of the Conference on the Human Dimension of the CSCE, 1990*, reprinted in 29 *International Legal Materials* (1990), available at http://www.osce.org/docs/english/1990-1999/hd/cope90e.htm (Washington, D.C.: American Society of International Law, 1990), 1305.

12. Initially conceived as a means to give aid to emerging democracies in the early stages of constitution-making, the Commission has taken on a range of subjects, including constitutional courts, national minorities, election laws, and other issues. See the Venice Commission website, http://Venice.coe.int/site/interface/english.htm.

13. See Ralf Dahrendorf, *Reflections on the Revolution in Europe*, 86–107.

14. Robert Dahl, *On Democracy* (New Haven: Yale University Press, 1998), 37.

15. Herman Schwartz, *The Struggle for Constitutional Justice in Post Communist Europe* (Chicago: University of Chicago Press, 2002), 7.

16. Irena Grudzinska Gross, "When Polish Constitutionalism Began," 6 *East European Constitutional Review* (Chicago, IL.: Published for the Center for the Study of Constitutionalism in Eastern Europe at the University of Chicago Law School in partnership with the Central European University, 1997). Indeed, one can make the case that reformers in the new era were especially drawn to declarations of rights because, unlike more tangible issues like wages or working conditions, they might not seem to entail officials' accountability to voters.

17. Rett R. Ludwikowski, *Constitution-Making in the Region of Former Soviet Domination* (Durham, N.C.: Duke University Press, 1996), 182.

18. Gross, "When Polish Constitutionalism Began," 74.

19. Article 3.

20. Cass R. Sunstein, "A Constitutional Anomaly in the Czech Republic?," 4 *East European Constitutional Review* (1995), 50, 51.

21. Ibid.

22. See symposium, "Enlargement as Seen from the East," 9 *East European Constitutional Review* (2000), 62.

23. Alina Mungui-Pippidi, "Romania," 9 *East European Constitutional Review* (1997), 77.

24. I develop this point in the section on "A Culture of Rights."

25. See Rett Ludwikowski, "Fundamental Constitutional Rights in the New Constitutions of Eastern and Central Europe," 3 *Cardozo Journal of International and Comparative Law* (New York, N.Y.: Benjamin N. Cardozo School of Law, Yeshiva University, 1995), 73, 88–90.

26. Amended Constitution, Article VIII, Section 1.

27. Gabor Halamai, "Democracy versus Constitutionalism? The Re-establishment of the Rule of Law in Hungary," 1 *Journal of Constitutional Law in Eastern and Central Europe* (Boxtel, The Netherlands: TFLR-Institute, 1994), 6, 13.

28. Article 31.3. See Gross, "When Polish Constitutionalism Began," 73.

29. Halamai, "Democracy versus Constitutionalism?" 15.

30. Ibid.

31. Mark Brzezinski, *The Struggle for Constitutionalism in Poland* (New York, N.Y.: St. Martin's Press; Oxford: In association with St. Antony's College, 1998), 201.

32. See Ibid., 202.

33. Judgment of June 24, 1994 (Defamation Case), translated and reprinted in 3 *East European Case Reporter of Constitutional Law* (Den Bosch, The Netherlands: BookWorld Publications, 1996), 148.

34. See Schwartz, *The Struggle for Constitutional Justice in Post Communist Europe*, 98.

35. *New York Times v. Sullivan*, 376 U.S. 254 (1964).

36. See Schwartz, *The Struggle for Constitutional Justice in Post Communist Europe*, 224–25. Indeed, the European Court of Human Rights, in Strasbourg, reversed a decision by a Romanian provincial court whereby a journalist was convicted for libel under Articles 205–06 of the Criminal Code. The minister of justice then instructed Romanian judges to interpret the Criminal Code's articles on defamation of authorities and libel with the understanding that international legislation prevails over national law in matters of human rights and press freedom (Article 20 of the Constitution). Constitution Watch, 8 *East European Constitutional Review* (1999), 2, 38.

37. See Cass R. Sunstein, "Something Old, Something New," 1 *East European Constitutional Review* (1997), 18.

38. See E.V. Niemeyer, Jr., *Revolution at Queretaro: The Mexican Constitutional Convention of 1916–1917* (Austin: Published for the Institute of Latin American Studies by the University of Texas Press, 1974), Ch. 4.

39. See chapter II of the Weimar Constitution, reprinted in Howard Lee McBain and Lindsay Rogers, eds., *The New Constitutions of Europe* (Garden City, N.Y.: Doubleday, Page, 1922), 197–209. As Gerhard Casper puts it, "After WWI, the Weimar Constitution of 1919, the most celebrated constitutional document of its time, attempted to constitutionalize the welfare state. Part 2 spelled out 'basic rights and basic duties.' It was subdivided into sections entitled 'The Individual,' 'The Life of the Community,' 'Religion and Religious Corporations,' 'Education and Schools,' and 'Economic Life.' There was hardly an aspect of life for which the constitution did not include governing principles of a programmatic nature, placing it all under the overarching care of the state. It reads at times as if it were the democratic version of Frederick the Great's General Code, and it left little in doubt that the regulatory powers of the state were comprehensive indeed." Gerhard Casper, "Changing Concepts of Constitutionalism: 18th to 20th Century," 1989 *Supreme Court Review* (Garden City, N.Y.: Doubleday, Page, 1922), 311, 327.

40. Hungarian Constitution, Article 70/B, Section 4.

41. Polish Constitution, Articles 65, 66.

42. See Czech Charter, Article 28.

43. Slovak Constitution, Article 36. As Jean-Marie Henckaerts and Stephaan van der Jeught point out, one "of the conflicts that caused the dissolution of Czechoslovakia was determining the appropriate degree of social protection." "Human Rights Protection under the New Constitutions of Central Europe," 20 *Loyola Los Angeles International and Comparative Law Journal* (Los Angeles: Loyola Law School, 2000), 455, 492–93.

44. Article 70/E.

45. See, e.g., Hungarian Constitution, Article 70/F; Polish Constitution, Article 70; Czech Charter, Article 33; Slovak Constitution, Article 42.

46. Czech Charter, Article 35, Sections 1, 2. Moreover, both the Czech and Slovak environmental rights come with companion duties on the part of citizens to protect, and, in Slovakia, even improve, the environment. See Czech Charter, Article 35, Section 3; Slovak Constitution, Article 44, Section 2.

47. Article 45.

48. Despite Article 45's categorical language, there has been some debate in the Irish courts as to whether the article may nevertheless be taken into account in deciding whether a claimed constitutional right exists. See J.M. Kelly, *The Irish Constitution*, Gerard Hogan and Gerry White eds. 3d ed. (Dublin: Butterworths, 1994), 1119–23.

49. The Directive Principles are set out in Part IV of the Indian Constitution (Articles 36–51). Article 37 states that these Principles are not justiciable; a court may not enforce them, but they are nevertheless to be considered "fundamental in the governance of the country." For a discussion of the drafting of the Directive Principles, see Granville Austin, *The Indian Constitution: Cornerstone of a Nation* (Oxford: Clarendon Press, 1996), Ch. 3.

50. Peter Paczolay, "Human Rights and Minorities in Hungary," 3 *Journal of Constitutional Law in Eastern and Central Europe* (1996), 111, 120.

51. See Czech Charter, Article 41, Section 1; Slovak Constitution, Article 51.

52. Henckaerts and van der Jeught, "Human Rights Protection under the New Constitutions of Central Europe," 491.

53. Polish Constitution, Articles 65, Section 5, and 66.

54. Article 70/E.

55. Paczolay, "Human Rights and Minorities in Hungary," 120.

56. Decision No. 31/1990 (XII.18.) AB h. of the Constitutional Court.

57. See Schwartz, *The Struggle for Constitutional Justice in Post Communist Europe*, 92.

58. Ibid., 65.

59. George Barany, "Hungary: From Aristocratic to Proletarian Nationalism," *Nationalism in Eastern Europe*, in Peter F. Sugar and Ivo J. Lederer, eds. (Seattle: University of Washington Press, 1969), 272.

60. Article IV.

61. Article 35.

62. Article 68.

63. Article 70/A.

64. Article 70/A, Section 3.

65. Peter Paczolay, "Human Rights and Minorities in Hungary," 111, 122.

66. See Ibid., 123.

67. Ibid., 124.

68. Ibid., 125.

69. On the problems of fostering vigorous local government in post-communist societies, see A.E. Dick Howard, "After Communism: Devolution on Central and Eastern Europe," 661.

70. Constitution Watch, 10 *East European Constitutional Review* (1997), 2, 20.

71. Ibid.

72. Henckaerts and Van der Jeught, "Human Rights Protection under the New Constitutions of Central Europe," 499.

73. Article 34.
74. Article 34, Section 3.
75. Henckaerts and Van der Jeught, "Human Rights Protection under the New Constitutions of Central Europe," 499.
76. Ludwikowski, "Fundamental Constitutional Rights in the New Constitutions of Eastern and Central Europe," 135.
77. Henckaerts and Van der Jeught, "Human Rights Protection under the New Constitutions of Central Europe," 500.
78. Constitution Watch, *East European Constitutional Review* (1994), 2, 21–22.
79. See Ludwikowski, "Fundamental Constitutional Rights in the New Constitutions of Central and Eastern Europe," 115.
80. Article IX (4).
81. Article IX (4); Article X (4).
82. Ludwikowski, "Fundamental Constitutional Rights in the New Constitutions of Central and Eastern Europe," 115.
83. An international poll released in September of 2000 indicated that 79 percent of Slovaks have a negative view of Roma, 46 percent believe that too many non-Slovaks reside in the Slovak Republic, and only 54 percent believe that Roma should have the same rights as Slovaks. See *U.S. State Department: Country Reports on Human Rights Practices—2000* (Washington, D.C.: U.S. G.P.O., 2000).
84. Ibid.
85. Ibid.
86. Quoted in Schwartz, *The Struggle for Constitutional Justice in Post Communist Europe*, 51.
87. Articles 79, 191 (6).
88. Article 190 (1) states: "Judgments of the Constitutional Tribunal shall be of universally binding application and shall be final."
89. See Kim Lane Scheppele, "Limitations on Fundamental Rights: Comparing Hungarian and American Constitutional Jurisprudence," 8 *Journal of Constitutional Law in Eastern and Central Europe* (2001), 43, 61–62.
90. László Sólyom, "The Hungarian Constitutional Court and Social Change," 19 *Yale Journal of International Law* (New Haven: Yale Law School, 1994), 223, 227.
91. Ibid.
92. Schwartz, *The Struggle for Constitutional Justice in Post Communist Europe*, 215.
93. László Sólyom and Georg Brunner, *Constitutional Judiciary in a New Democracy: The Hungarian Constitutional Court* (Ann Arbor: University of Michigan Press, 2000), 38.
94. On legal certainty and principled coherence as pillars of the rule of law in Hungary, see Ibid., 38–42.
95. Thomas Garrigue Masaryk, *The Making of a State: Memories and Observations, 1914–1918* (New York: Frederick A. Stokes Company, 1927), 464, 465.
96. See Dahrendorf, *Reflections on the Revolution in Europe*, 86–107.
97. Ewa Letowska, "A Constitution of Possibilities," 6 *Eastern European Constitutional Review* (1997), 76.
98. Andras Sajo, "Universal Rights, Missionaries, Converts, and 'Local Savages,'" 6 *East European Constitutional Review* (1997), 44, 45–46.

99. See A.E. Dick Howard, *The Road from Runnymede: Magna Carta and Constitutionalism in America* (Charlottesville, VA: University of Virginia Press, 1968), 135–202.

100. Gross, "When Polish Constitutionalism Began," 69.

101. Sajo, "Universal Rights, Missionaries, Converts, and 'Local Savages,' " 46.

102. Letowska, "A Constitution of Possibilities," 80.

103. Sajo, "Universal Rights, Missionaries, Converts, and 'Local Savages,' " 48–49.

104. Scheppelle, "Limitations on Fundamental Rights: Comparing Hungarian and American Constitutional Jurisprudence," 65.

105. *Notes on the State of Virginia*, ed. William Peden (Chapel Hill: University of North Carolina Press, 1955), 148.

106. Section 15.

PART III

CASE STUDIES AND THE IMPLEMENTATION OF RIGHTS AND DEMOCRACY

CHAPTER 9

THOMAS JEFFERSON, RIGHTS, AND THE
CONTEMPORARY WORLD

Yasushi Akashi

The impact of Thomas Jefferson's political ideas on the world community is far-reaching, particularly his ideas on rights and democracy. In the 1950s, I was a youthful Jeffersonian student in Japan, writing a bachelor's thesis on his political thought. Then, I studied at the University of Virginia, a fruit of Jefferson's dreams, as a Fulbright scholar. Subsequently as a senior United Nations official, I traveled far and wide in the 1990s, dealing with ethnic conflicts and humanitarian crises. As an Asian, I cannot help but appreciate the complexity of putting into practice human rights, and more particularly ways and means of implementing these rights in the diverse contexts of the Asian environment.

As a student brought up in postwar Japan, which was under the American occupation after its defeat in World War II, I was keenly interested in exploring Jefferson's ideas, which often seemed to lie behind reform policies carried out by the Occupation authorities. Through extensive reading, I found that Jefferson was more a pragmatic political leader than a scholar, purely interested in the consistency of his thoughts. For instance, in the course of his life, Jefferson expressed divergent, sometimes contradictory, ideas about religion. While respecting established religions, Jefferson appears to have been essentially an agnostic, with an open, liberal, and skeptical mind.

The contradiction between his theory and practice was particularly evident when he confronted the question of purchasing Louisiana from France so as to make the United States a powerful nation, more than twice as large as before in territory. To purchase such a huge territory was contrary to his constitutional view, based on a strict construction of the Constitution. In the end, he decided in favor of what he thought a broader national interest, rather than being faithful to his cherished interpretation of the Constitution.

Jefferson's concept of democracy was not an abstract thought. His idea of the best functioning of democracy was when it fostered the independent farmer endorsed with his own economic sustenance. Jefferson was almost idyllic whenever he described American democracy based on self-reliant and independent farmers. In my view, his ideas of democracy were distinctly different from those of many of his colleagues like Alexander Hamilton and James Madison who were thinking of the stability and prosperity of the American political system on the basis of emerging entrepreneurs and workers.

Jefferson, in my opinion, was one of a small minority of thinkers of his time who believed in the goodness of human nature, while many others like Hamilton and Madison were more concerned about how to control its selfish passions and the excesses inherent in popular democracy, which could undermine the integrity and balance of the constitutional system. Without doubt Jefferson exerted the greatest influence on the drafting of the Declaration of Independence. But his colleagues such as Hamilton and Madison were far more influential with regard to the drafting of the Federal Constitution in Philadelphia, as can be vividly perceived from Max Farrand's voluminous work. It was their deep-seated mistrust of popular democracy that made them insist on the strict separation of powers among the three branches of government, and advocate indirect, rather than direct, election of the President as well as of the Senators. One can only speculate how differently the American Constitution might have been framed had Thomas Jefferson been present at the drafting exercise in Philadelphia.

The concept of human rights in modern history has evolved over the last 300 years. The American War of Independence and the French Revolution were two epochal developments in the course of that evolution. Before the establishment of the United Nations in 1945, human rights were considered to belong essentially to the realm of domestic jurisdiction by sovereign governments. Consequently, expressing concerns in the human rights situations of other countries was felt to constitute interference in internal affairs of these countries. This thought is embodied in Article 2, Paragraph 7 of the UN Charter, which states:

> Nothing contained in the present Charter shall authorize the United Nations to intervene in matters which are essentially within the domestic jurisdiction of any state or shall require the Members to submit such matters to settlement under the present Charter; but this principle shall not prejudice the application of enforcement measures under Chapter VII.

At the same time, however, the framers of the UN Charter thought that war constituted the most grievous violation of human rights and that, in order to establish genuine international peace, it was vital to guarantee human rights on an international basis. Thus the Charter makes clear that the achievement of

> international cooperation in solving international problems of an economic, social, cultural or humanitarian character, and in promoting and encouraging

> respect for human rights and for fundamental freedoms for all without distinction
> as to race, sex, language, or religion

is one of the Organization's principal purposes.

The Universal Declaration of Human Rights adopted by the UN General Assembly in 1948, was a landmark based on the idea that there is no peace where there is no human freedom, and humanity cannot be free when there is war and the threat of war. In 1966 the Universal Declaration was codified into legally binding form in the two Covenants dealing, respectively, with Economic, Social, and Cultural Rights and with Civil and Political Rights.

In the face of the reality prevailing in Asia and against the background of its cultural and political traditions, human rights have been a subject of great debate. There will be no easy answer to the question of democracy and human rights in the developing countries of Asia, Africa, and Latin America, which are coping with serious geographical and economic conditions and the pressures of population explosion. In these countries, the relationship between the individual and his society is in constant flux. The debate on individual rights and group rights has not been settled, although indications are that individual rights are being asserted with greater vehemence and conviction today. The equilibrium between freedom and responsibility has also baffled philosophers. Former prime minister of Singapore Lee Kuan Yew created a controversy by comparing the Chinese and the Russian approaches toward modernization, and by expressing his preference for the Chinese approach, which emphasizes economic and industrial growth of the nation as compared to the Russian system, which emphasized individual political rights even before economic preconditions were met.

After completing the ambitious tasks on behalf of the United Nations in Cambodia with the adoption of a democratic constitution by the Cambodian people, I stated in September 1993 as the head of the United Nations peace-keeping operation in that country that, although Cambodia faced an enormous challenge in its future, I did not believe that development and democracy were incompatible with each other or that human rights and the unity of state could not be harmonized with each other. In my view, to meet these two seemingly contradictory challenges was within the reach of Cambodians. Obviously, for a developing country, which did not yet possess satisfactory material conditions or a sufficiently developed middle class, the challenge was enormous, but I expressed an optimistic view that this was not an unattainable aim for the Cambodian people. It is hoped that the subsequent developments in Cambodia have borne out the accuracy of my belief, despite occasional political crises the country went through.

At the World Conference on Human Rights, held in Vienna in 1993, the parallel viewpoints, held by the developed North and the developing South, became evident. Many of those who were from newly independent states wanted to stick to traditional state sovereignty, and many of those who wanted to ensure human rights and rights of citizens were from developed countries. Ultimately the conference recommended the establishment of the post of the United Nations Human Rights High Commissioner, which has in fact taken place. The final document adopted at the Vienna Conference states that human rights and basic freedoms are universal. The states are obligated to protect these rights, irrespective of political, economic, and cultural systems. However, as a gesture of compromise, the final document refers to the importance to be given to national and regional characteristics and diversity in historical, cultural, and religious backgrounds. The right of development is also considered universal and inalienable, and it constitutes a part and parcel of basic human rights today.

Is democracy essentially a value to be cherished, or political procedure to be followed? It is both, but it may more be a vital value and indispensable goal than a matter of political procedure. The notion that all individuals have equal dignity is the cornerstone of modern society. It is of interest in this connection that the concept of "human security" has become an important principle in contemporary international relations, supplementing the traditional "national security."

Modern democracy is often considered to be a system of representation, based on decision by majority, but it should not be forgotten that a minority in a political community possesses certain inviolable rights within that community. The ethnic conflict in the Former Yugoslavia, in which I was deeply involved on the side of the United Nations, was sparked partly by the hastiness with which some European governments recognized the governments of Croatia and Bosnia, before these governments promised to ensure the rights of minority groups in their constitutions, as recommended by the Bardenter Commission. This leads me to the reflection that one painful lesson from ethnic conflicts of the 1990s is that real democracy must contain, as an essential component, full respect to be paid to all individuals and minorities—ethnic, cultural, or religion- or gender-based.

We continue to be inspired by the brilliant insight and ringing admonitions on human rights offered by Thomas Jefferson, who possessed unflinching faith in human rights at a time when such a thought was still considered dangerous. At the same time, we have to admit the economic basis for Jeffersonian democracy, whose roots were intimately based on the independent farmers of the agrarian society of his time.

Democracy and human rights have become an inevitable wave of history, transforming all countries and regions of the world. The right of self-determination of each people to form their particular form of government and choose the suitable pace of realization of their ideals in this regard must be fully accepted. In doing so, we should avoid moral arrogance or hasty attempts to impose one's own particular system on the rest of the world. Whether the international community has the right or obligation to intervene in humanitarian crises in various parts of the world simply cannot be decided in the abstract. Such decisions will have to be made most carefully, case by case, taking into account all the elements each circumstance of the situation contains.

CHAPTER 10

CITIZENSHIP AND THE STRUGGLE FOR RIGHTS IN FLEDGLING DEMOCRACIES

Mark Tamthai

> *. . . . A good citizen is one who acts according to the dictates of the government, since the government knows what is best for them and for the country.*
>
> *Duties of Citizens,* 4th grade text, 1960 Ministry of Education
> (Thailand)

The struggles for rights that are taking place in the new democracies of today's world have features both similar and different from those that took place in the world's new democracies in the latter years of the eighteenth century. Where they are different is, because of the successes of the earlier struggles there now exist international conventions and declarations protecting these rights, which a significant portion of the world adheres to. The norms these present struggles aim to establish in their countries are ones that are accepted elsewhere, and thus the struggles are not as local as the earlier ones. The language of rights used in the discourse of these contemporary struggles is the same wherever they are taking place. This last feature has led some to question whether or not these struggles are truly those arising out of particular local situations by persons who feel their rights are being violated, or by another group on their behalf. And if the latter, then there is a concern that the struggles will not be able to sustain themselves. However, also taking place at the same time in these countries are struggles much more similar to the eighteenth- and nineteenth-century struggles. They are being undertaken by people in these countries who feel that they have continually been taken advantage of and not given a place in society, and who are saying that they will no longer accept such a situation. Their feelings are expressed in their own language and make reference to their own cultural predicament. Instead of saying, "My rights in accordance with the Universal

Declaration of Human Rights are being violated!" they might say something like, "You don't see my head and I won't take it anymore!." Neither one of these two kinds of struggle is more legitimate than the other. The fact that there exist two such kinds of struggles merely reflects historical realities. However, even if these two kinds of struggles, which are taking place (in the same country) today, do seem to have important differences, what they have in common is that they both have to overcome the same traditional ideas and practices that are obstacles to establishing a culture of rights in their particular country. Though one kind of struggle could be seen as a "top down approach" and the other a "from the bottom up approach," they support each other in their struggles. The intention of this essay is to address the question of the role of citizens in both of these kinds of struggle for rights, which are taking place in fledgling democracies.

We use "fledgling democracies" here to mean those countries that have just "recently" begun their experiment in democratic government. This could mean a couple of years, or a much longer time if it was continually punctuated by periods of authoritarian rule, such as the past 70 years in Thailand. In addition to the length of time these countries have undertaken the democratic experiment, the fledgling democracies we consider also have two further characteristics. The first characteristic is that their experiment with democracy began in the midst of a number of structural injustices. These could be land issues, the status of minorities, wide spread poverty, or other such problems. The second characteristic is that of having a long history of being a hierarchical society. This social structure could have come about from previous forms of authoritarian government, such as an absolute monarchy, or from other cultural factors, such as having a caste system. This specific scope of the kinds of countries considered in this chapter may not cover all fledgling democracies in the world today, but covers enough, I believe, to merit our attention.

The two characteristics of fledgling democracies mentioned above present a unique problem to the question of the protection of rights in these societies, and the role of citizens in this endeavor. In the first case, the existing injustices often have roots that go so deep and are so encompassing that it is sometimes claimed that exercising citizenship in these cases actually prolong these injustices. If this is indeed the case, then in some perverse way democracy, rather than protecting rights, becomes an obstacle to remedying cases of rights violations. As for the second case, the problem arises from the fact that though democracies in themselves are supposed to protect rights by various mechanisms, these mechanisms depend for a great part on the people taking citizenship seriously. However, these demands of citizenship often come up against various others imposed by a long history of social hierarchy, and finding a way over, around, or through these kinds of obstacles is one of the main challenges such societies face. Besides examining individually the details of these two kinds of challenges, we also try to see how they are related and propose some possible ways to meet them. The last section takes up the question of the relevance of the ideas of Jefferson, or Jeffersonian scholarship, to the problems mentioned above.

Throughout the chapter different points are illustrated by using historical and cultural examples from Thailand. Though this is done by necessity since these are the examples I am most familiar with, I believe for the most part they can be substituted appropriately by the particularities of other fledgling democracies to result in similar conclusions.

CITIZENSHIP, THE PROTECTION OF RIGHTS AND STRUCTURAL INJUSTICES

Many fledgling democracies come into being as a result of a struggle for the holding of "free and fair" national elections. Having accomplished this, with varying degrees of success, representative democracy then goes in to full swing. However, since the underlying sentiment of what elections should be about lie very close to a form of aggregate democracy where the preferences of the people are added according to a particular system without regard to how they came about, the pre-election status quo with regard to the structural inequalities tends to remain in place after the votes are counted. Elected representatives then not only continue to oppose movements that try to correct the rights violations occurring in society, but now do so on the basis of their being the legitimate representatives reflecting the will of the majority. Since these duly elected officials mirror the social prejudices in existence, in a large number of cases their claim of reflecting the will of the majority may not be that far from the truth. The source of the problem seems to be the understanding of the kind of democracy that is being developed. As long as democracy is understood merely as "non-dictatorship" and the tallying of people's views is seen as being adequate to bringing about "rule by the people," then this situation will not only continue to remain but will grow roots even harder to dig out. The struggles against injustice, which have often gone hand in hand with the fight for democracy, will then lose one of its more potent weapons. The remedy is to instill a different understanding of democracy in such countries, and the idea of democracy that would begin to address this problem needs to be along the lines of deliberative democracy, which I take as having the following main features. A deliberative democracy is one where the people (a) deliberate and vote their informed judgments as to what is for the common good, (b) recognize that because of the possibility of different conceptions of "the common good" it is their responsibility to cite public reasons for their views, and (c) agree that such public reasons need to be acceptable to others in their capacity as democratic citizens and not based on any particular religious or philosophical doctrine. Such an understanding of deliberative democracy is similar I believe to what has been proposed by others (see e.g., Freeman 2000). Though there has been much debate in recent literature as to the problems faced by such an understanding of democracy, together with various defenses and justifications for making deliberation a cornerstone of democratic citizenship (see e.g., Freeman 2000; Dryzek 2001), the question of whether this is the appropriate understanding of democracy in the context of fledgling

democracies is different from raising this same question in the usual contexts
in which the debates have taken place. Those contexts are usually of apathetic
voters not taking much interest in public matters. Two objections to deliberative democracy in such contexts are that if deliberation is the legitimizing
factor then in the real world of these democracies most decisions made are
not legitimate, and second, it is impractical if not impossible to require large
scale deliberation for reasons of economy. The first objection seems to be a
case of begging the question whereas the second objection confuses widespread deliberation as an "ideal to approach ever closer" and an actual measure of whether a system of government is democratic or not. For fledgling
democracies though, the real value of public deliberation lies in the fact that
it is one of the best ways to allow pre-existing social prejudices to be brought
up to the surface for acknowledgement and scrutiny. This can then possibly
lead to real attempts to correct the structural injustices of society. However,
I say possibly because though public deliberation may be a necessary condition in fledgling democracies for the eventual protection of rights it is not a
sufficient condition. What is still needed to bring into effective play all the
features of deliberative democracy is some kind of good will between citizens
(see Tamthai 1999 for a further elaboration on the need for good will
between citizens in fledgling democracies). Without some kind of common
concern between citizens true "public" reason will not be used. If people different from you are seen as "others" there will be a tendency to see "your
reason" as "public reason" (this is similar to one of the points made by
Young 2001 concerning the objections raised by activists toward putting
deliberation into practice prematurely because of the problem of hegemony).
If there is no accepted idea of equality present liberty becomes part of the
problem and not part of the solution. The way to proceed here is to concentrate on building a sense of fraternity in society. Liberty together with fraternity does have a chance to succeed in addressing the inequalities that exist.

Promoting fraternity in fledgling democracies requires studying what historical and cultural resources exist that might be used toward this end. For
the case of Thailand there are two different types of resources. The first
might be more difficult to utilize since it relies on a moral system, Buddhism,
which though accepted as an undisputed ideal is more difficult to put into
practice. On the positive side is that even small steps toward this ideal will go
a long way toward building a long-lasting sense of fraternity. The second
resource, the present monarch, can have immediate effect but may not be
sustainable since reigns change.

The Bhuddhist dhamma most closely related to building a sense of fraternity in society is the Saraniyadhamma (see Payutto 1994 for a particular view
on this matter). The Saraniyadhamma, or Six Conditions for Fraternal Living,
is a teaching meant for creating harmony and cohesion throughout a society
(see dhamma 259 in Payutto 1984). The six conditions are as follows:

> mettakayakamma; to show good will towards others in deeds
> mettavacikamma; to show good will towards others in words

mettamanokamma; to show good will towards others in thoughts
sadharanabhogi; to share rightfully acquired gains with others, whether they
have less or more
silasamannata; to be honest towards others and live according to moral
precepts
ditthisamannata; to have similar fundamental ideals, such as on how people
should live together.

The first five conditions are prescriptions individuals can work toward
whereas the sixth condition seems to be more of a description of the kind of
communities where this dhamma would be effective. Though, as was mentioned above, it is a long road to take if one hopes to make changes on the
basis of religious teachings, in Thailand using dhammas to question policies
and actions that violate basic rights can prove to be effective if a proper place
and method to raise such matters can be found.

The case where the present King can play a role in creating fraternity (see
Tamthai 2000 for a detailed proposal along these lines) is very time specific,
yet similar conditions might be duplicated elsewhere or in other times. The
main idea here is for the King to be the central "passing-on-concern" point
in society. The contention is that "caring about," under certain conditions of
knowledge, is a transitive relation. If we care about someone, and that person cares about another person, *and we know this*, then we will care about
that other person as well. In the situation we are considering, if everyone
loves the King, and the King shows publicly that he is concerned with the
welfare of all members of society, then everyone will share in this concern as
well. If this analysis is correct, then a very real question that needs to be
answered in the case of Thailand is why the sense of fraternity in Thai society
is not as widespread as it should be. The answer is that the fundamental
importance of creating fraternity has not yet been elaborated on enough, and
therefore there has not been any attempt beyond the usual efforts to show
publicly the King's concern for the welfare of the people of Thailand.
Because of the wide range of injustices what is required is constant detailed
affirmation to the public of his concern for *all* the people. Even if this is
brought about though, there will still be the problem of the short-lived
nature of this solution. It can work because the people of Thailand really do
love and respect the King. Such a condition need not always be the case,
however, as history has shown.

Both these approaches to building fraternity share a common feature,
which I think is essential to making a practical difference in today's fledgling
democracies. This feature is that of relying on inter-relatedness to provide
the necessary mutual concern in society rather than on shared memories or
past glories. An actual example of a consequence of such a feature is the new
national security policy for the southern border provinces of Thailand, which
instead of being based on the idea of assimilation as in the past is based on
the idea of a shared future with all citizens contributing toward building a
central wisdom. The challenge of creating a sense of fraternity is not dissimilar

to the challenge of creating a sense of national identity. The world is as we find it, and whatever the past legacies might be in terms of the origin of present geographical borders there is not much likelihood of changes occurring in the near future. Meanwhile violations of fundamental rights continue, and new ways must be found to deal with them.

CITIZENSHIP AND THE PROTECTION OF RIGHTS IN HIERARCHICAL SOCIETIES

Supposing that in the fledgling democracies there has been created a certain level of fraternal feeling among citizens, and that basic rights have not only been enshrined in documents such as constitutions, but have been accepted in the thinking of the members of society as well. The right to exercise reason and freedom of opinion are now accepted norms of society. With this safeguard in place the foundations for a functioning deliberative democracy can now begin to be put in place. What are the immediate challenges that need to be met in such situations to provide this foundation? Certainly foremost among them is to develop the ability of citizens to be able to exercise their right to take part in public debate in an informed manner. This is a task to be handled by formal and non-formal educational institutions. But besides *having* the right and freedom to engage in debate on public issues, and *being able* to do so in an informed manner, there is still missing an essential piece to this puzzle, and that is the *desire* to participate in such activities. This last piece that needs to be added in order for citizens to really begin to take part in deliberative democracy is probably the most difficult to develop. It is also essential for the protection of rights because it is when citizens of all backgrounds engage in public deliberation that individual cases of violations of rights that may have been overlooked are brought to light. The reason that it is so difficult to develop is because it requires individuals to overcome some very difficult psychological barriers erected by the hierarchical systems that have existed in the society for hundreds, or in some cases even thousands, of years. One such barrier is the culturally entrenched ideas of what a "good person" or a "cultured person" is, which are at odds with the idea of what a "good citizen" in a deliberative democracy is supposed to be like. Another barrier is the lack of belief in one's ability to contribute as fruitfully as those who have been the historical decision makers, toward deliberation on public matters. We will consider examples of each kind of barrier in the Thai context.

Sombat kong Phu Dee (Chao Phraya 1997) is a book that has probably had as much influence on Thai society's formation of what the characteristics of a "cultured" or "good" person consist of, as any other book or source of teachings. The reasons for this are twofold. First, it is a book that has been widely disseminated directly and indirectly. First published about 100 years ago it is now in its twenty printing with the total number of issues exceeding one million. Its main use has been in the classroom at the primary school level, which is part of Thailand's compulsory education. This means that most Thai people of today have had to read it. For those who have not been directly exposed to it, the lessons in the book are passed on to them via their many interactions with members of society who are familiar with it. *Sombat kong Phu Dee* is a small book that lists 182 things a "cultured" person should or should not do. These 182 prescriptions are divided into 10 sections, with the items in each section grouped according to physical behavior, verbal behavior, and thought behavior. The first two groupings can of course be easily monitored, and for young children in Thailand leads to a kind of conditioning. The third group of thought behaviors is more problematical to monitor and, consequently, such kinds of behavior have not played a big role. Some of these prescriptions describe what Thailand considers to be good individual manners. Here are some examples.

- *Do not stand higher than a person who is older than you. If such a person is sitting down, then you should sit or kneel.*
- *Do not touch another person unless that person is a friend or a family member.*
- *Do not reach across other people to get something.*
- *Do not interrupt when other people are speaking.*
- *Always speak in a soft voice, not showing your emotions loudly.*
- *Do not speak about yourself or your personal matters in public.*

However, besides these items of etiquette there are also a number of prescriptions having a direct influence on public behavior. For example:

- *Do not speak critically or sarcastically in public of "pu yai" (those people of high social status).*
- *Do not walk in front of "pu yai," always follow behind.*
- *Do not embarrass your teachers or elders.*
- *Do not criticize or ridicule other people's ideas in public.*

What the latter group of examples does is create in the minds of Thai people a type of desirable behavior that at times is a direct opposite to the kinds of behavior exhibited in deliberation and debate of public issues. To reconcile this with campaign behavior before elections, or even behavior in parliament, there has developed in Thai society an acceptance that politicians are somehow exempt from these rules due to their "profession" but that for people in general they still hold. This makes it difficult to encourage public participation because for some people it is almost like becoming "non-Thai." Democratic deliberation in Thailand then comes close to being a self-contradiction, since public reason in this context leads to the conclusion that public debate is not desirable. This is why the promotion of deliberative democracy may require something close to a cultural revolution. Such a revolution though is not alien to certain ways of thinking that are accepted as an ideal though not exhibited clearly in practice. A very relevant example here is the Buddhist dhamma Kalamasuttra (Dhamma 305 in Payutto 1985), which teaches one not to accept prescriptions just because they are taught by those one respects, or are printed in widely read books, but to accept them after experimenting to see if they bring happiness to the individual and to society at large. If this direction is taken then what needs to be done might resemble cultural reform more than revolution.

The second barrier alluded to earlier is the lack of self-confidence that is found in most people as to how much value their ideas really have and that therefore it would be better if debate is left to those who "know better." This is a remnant of the Sakdina system in Thailand, which was operative in the past for a period of at least 600 years. The Sakdina system was a social hierarchy in place officially until early twentieth century. It was the cement holding together a legal system codified by King Rama I a little over 200 years ago (see *Three Seal Code of Laws*, 1978). This legal code of more than 1,200 articles consisted of laws relating to treason, criminal matters, role of witnesses in judicial proceedings, punishment, taxation, hereditary issues, civil disputes, and the Like. The essence of the Sakdina system is that different members of society have different social ranks, indicated numerically by a unit referred to as "na," depending on their birthright, position held and type of work. Most of the laws in the *Three Seal Code* make reference to the "nas" of the people involved in the particular cases. For example, (a) how much weight is given to a person's testimony in a trial depends on the witness' "na." The testimony of a person of higher "na" is given more weight proportionately than that of one of lower "na," (b) punishment for crimes depend on the "na" of the perpetrator and the "na" of the victim, (c) first wives have half the "na" of their husbands, second wives half of first wives, and so on, (d) if a person is accused of some matter and then acquitted, the accuser is punished according to the "na" of the one accused, greater punishment for accusing people of higher "na," (e) for some crimes which involve being knowledgeable of a particular matter, persons of lower "na" receive less punishment because they are considered not being able to understand such things.

These kinds of laws in effect resulted in the institutionalization of inequality of a very subtle and deep nature. People of low "na," which were the majority of people as in all feudal systems, came to believe that they really did have little worth. What developed was as intricate a patronage system as has ever existed. The fact that many patrons, or people of high "na," did not treat those below them too harshly resulted in the system becoming even more entrenched, and though officially this legal code is no longer in use its influence on how one understands one's place in society is still evident. When an individual hears a person who holds a position which in days past would be of higher "na" than themselves offer an opinion on a matter of public concern, the immediate reaction is that not only would it be a sign of disrespect to disagree publicly, but that such a disagreement is probably based on incomplete understanding. If that person is the King, then all public debate and deliberation dissolves. As with the previous barrier, encouraging deliberative democracy in such a context also requires cultural reform. Though the idea of cultural reform is distasteful and frightening to some parts of Thai society, it is actually a natural process of a large community trying to find the best way to live together. This reform will eventually be brought about by the "pu yai" of Thai society themselves, in some cases by consciously refining their own public deliberation to make it less intimidating, while in other cases expressing opinions which are subsequently shown clearly as being foolish or just plain wrong. The past five years have shown an increase in the number of latter cases and an accompanying increase of self-esteem, especially among the rural villagers.

THE RELEVANCE OF JEFFERSON FOR FLEDGLING DEMOCRACIES

To discuss this topic in depth one would need to be a Jeffersonian scholar and a student of late-eighteenth to early-nineteenth-century America. Presented here are merely a set of queries and some remarks.

(1) It has always been a point of discussion whether today's new democracies can learn lessons from the early periods of the more established democracies. For such to be the case it would require that the problems faced by each have enough similarities. Is the setting that Jefferson was in similar to the setting of fledgling democracies discussed in this chapter? In particular, were the two main characteristics of fledgling democracies, the existence of structural injustices and a hierarchical society, also characteristics of late-eighteenth to early-nineteenth-century America? Are there other important differences or similarities? That there were structural injustices in America at that time is without question. One needs go no further than the slave trade, the treatment of native Americans, and the status of women. To counter that these were not seen as such at the time would just be similar to saying that the injustices in today's fledgling democracies are not seen as such by certain members of these societies as well. So here we do find a similarity. As to the hierarchical nature of American society in that period, the answer seems to be ambiguous. Though slaves, native peoples, women, and white males occupied different levels of society, the group of white males that held most of the political power could be seen as more or less an egalitarian community. These egalitarian ideas were embraced purposefully when leaving more class-conscious societies in Europe to come to the new world. In this sense it is different from the social structure of fledgling democracies that exhibit a hierarchical nature throughout and is a legacy left over from the past, which must be dealt with.

(2) Did Jefferson attempt to deal with the injustices around him in a way that fledgling democracies today might find helpful in trying to meet this particular challenge? Here we run into the problem that has been referred to as Jefferson's dual nature. His ownership of slaves (for whatever reason, economic or otherwise) and his views on women make it difficult to see where his value to fledgling democracies would lie on

this issue. His behavior while in France happily participating in salon life discussing political matters with various women, while behaving in an opposite manner back in America (Marvick 1999), brings to mind the schizophrenic behavior of many members of Thai society who discuss and debate with each other in a different manner when in the company of friends from the West than when they meet among themselves. This refusal to examine one's own fundamental norms of public behavior is one of the obstacles fledgling democracies must deal with. Though it has been pointed out by many that Jefferson's ideas were eventually the main catalyst for dealing with these injustices, "eventually" in the case of America was a period of around 150 years. Today's fledgling democracies are not able to wait that long.

(3) One topic that is very relevant to the difficulties faced by fledgling democracies is the role of religion in attempts to meet these challenges. Here the examples from particular countries may not have counterparts since fledgling democracies are of many different faiths, Christian, Buddhist, Islamic and so on. Jefferson's various attempts at reconciling his religious and political views through a separation of church and state, while still holding on to the importance of virtue for citizenship, would be worth studying in detail. Since these attempts relied on the demystification of Christianity and the substitution of some of what is lost in the process by "the holy cause of freedom" (Sheridan 1999), whether a similar tract exists for other theistic religions needs to be explored. For fledgling democracies set in a non-theistic religious context, such as Buddhism, this critical analytic stance of enlightenment should theoretically be possible to accommodate. The main question to be answered here is whether or not deliberative democracy requires a shift from a theological/transcendental to a secular world view.

(4) Another set of ideas of Jefferson's that is intriguing to consider in the context of fledgling democracies, and which just might be very relevant, are his ideas on education and its role in developing future citizens. These ideas have not seemed to have sunk deep roots in America, possibly for similar reasons as why ideas on educational reform in Thailand are primarily geared toward meeting the challenges posed by global economic competition. But if globalization is seen in a broader framework that includes also matters of global justice and its potential for emancipation (see Tamthai 2002 for a proposal along these lines) then Jeffersonian ideas of education as citizenship building would be relevant to fledgling democracies, especially if education policies along these lines address directly the two fundamental challenges.

(5) In his discussions on education, does Jefferson speak of instilling a sense of fraternity among members of society as part of citizenship education? If so, much could be learned that would be of use to fledgling democracies today from a study of his prescriptions on this matter.

(6) The final point to consider is perhaps the most basic, that is, Jefferson's non-foundational view of democracy as perpetual revolution. Emerging

out of long authoritarian histories and steeped in traditions, "The Earth belongs always to the living" is just too radical a position for most fledgling democracies today. Even the most radical thinkers of fledgling democracies seem to be looking for long-term political stability, with only as many waves as is needed in a functioning deliberative democracy.

To conclude this section, one would have to say that studying too much of Jefferson would not be helpful to fledgling democracies in trying to meet their challenges of citizenship. However, some of his ideas could prove useful, and so these countries would need to undertake a careful process of identifying such ideas. Since Jefferson is not part of their histories, they can afford this luxury of selection, which would then make it easier to concentrate on ideas rather than personality.

NOTE

Presented at the conference "Thomas Jefferson, Rights and the Contemporary World" organized by the International Center for Jefferson Studies at Monticello, held at the Rockefeller Center in Bellagio, June 3–7, 2002.

REFERENCES

Chao Phraya Phra Sadejsurentaratibdi (1997) *Sombat kong Phu Dee* (Characteristics of a Cultured Person), 24th printing, Ministry of Education, Thailand.

Dryzek, John S. "Legitimacy and Economy in Deliberative Democracy." *Political Theory* vol. 29. (October 2001).

Freeman, Samual "Deliberative Democracy: A Sympathetic Comment." *Philosophy and Public Affairs* vol. 29. (Fall 2000).

Marvick, Elizabeth W. (1997) "Thomas Jefferson and the Old World: Personal Experience in the Formation of Early Republican Ideals" in James Gilreath ed., *Thomas Jefferson and the Education of a Citizen*, Washington: Library of Congress.

Ministry of Education (Thailand) (1960) *Duties of Citizens*. Division for 4th Grade Curriculum, Bangkok.

Payutto, P.A. (1985) *Dictionary of Buddhist Dhammas*. Bangkok: Mahachulalongkorn Rachawitayalai Press.

Payutto, P.A. (1994) *Buddhist Solutions for the Twenty-First Century*. Bangkok: The Bhuddhadhamma Foundation.

Sheridan, Eugene R. (1999) "Liberty and Virtue: Religion and Republicanism in Jeffersonian Thought" in James Gilreath ed., *Thomas Jefferson and the Education of a Citizen*.

Tamthai, Mark "Democracy with a Heart: Steps Towards a Wiser Millennium" in Philip Cam, In Suk Cha, Ramon Reyes, and Mark Tamthai eds., *Philosophy, Culture and Education—Asian Societies in Transition*, Korean National Commission for Unesco.

Tamthai, Mark (2000) "Garn Pokkrong Rabob Prachatipatai un mee Phra Maha Ksatra song pen Pramook" (On Constitutional Monarchies) in Jaran Kosananonda ed., *Withee Sangkom Thai* (Way of Thai Society), Pridi Phanomyong Institute, Bangkok.

Tamthai, Mark (2002) "Addressing the Democratic Deficit: Trickle-down Democracy and Learning to Walk the Deep Talk," the Second Asia-Europe Roundtable, September 2001. Oxford, UK: Corpus Christi College.

Three Seal Code of Laws (1978) 6th Edition, Division of Fine Arts, Ministry of Education, Thailand.

Young, Iris M. "Activist Challenges to Deliberative Democracy." *Political Theory* vol. 29 (October 2001).

CHAPTER 11

HUMAN RIGHTS OF WOMEN IN IRAN

Nesta Ramazani

Ever since the encounter of Muslim societies with modernity, no other area of human rights has been as resistant to change as family law, especially women's rights, under Islamic law (*Shari'a*). No matter what progress had been made with respect to women's rights before the Iranian Revolution of 1979, all human-made laws were subsequently subjected to Islamic law criteria, particularly those pertaining to women's rights. Yet, over the past quarter century no aspect of social life has remained immune to the dynamics of change, including the status of women.

The raging debate on human rights regime between universalists and cultural exclusivists granted, this essay suggests that many women's rights issues cannot and should not be squeezed into the abstract straitjacket of either school of thought because neither approach would adequately factor in the reality of evolutionary processes of change taking place—even in the Islamic Republic of Iran over the past quarter century. Empirically, the mixture of continuity and change is the hallmark of women's rights in Iran.[1] The pace of change in women's rights since the election of Mohammad Khatami to the presidency in 1997 has somewhat accelerated.

With a strong commitment to incrementally moving women's rights toward more universal norms, he has single-mindedly supported the enhancement of such rights despite dogged resistance on the part of hardline conservative factions within and outside the establishment. Generally speaking, Khatami seems to reject the binary stance juxtaposing universal and cultural standards and norms. The Iranian ambassador to the United Nations, for example, told the General Assembly on December 11, 1998 that "...the fact of the matter is that global pluralism and cultural diversity on the one hand and universality of human rights on the other are not contradictory, rather, they are mutually reinforcing..."[2] Six years later, however, this formulation does not seem to satisfy reformers who have been disappointed with the snail-like pace of progress in women's rights despite President Khatami's enlightened vision on human rights. The pressure for moving more rapidly towards universal norms is increasing.

On the basis of documentary research and on-site interviews with scores of women activists, I have selected a variety of issue areas regarding women's rights that best demonstrate this fascinating process of continuity and change, resulting in an opaque mix of universalist and cultural relativist norms and practices. At the end of the essay, I try to speculate briefly about the future prospects for improving women's rights with an eye to both domestic and external pressures that seem to drive women's rights increasingly away from exclusively cultural encrustations toward a brighter future of universalist ideals. On balance, however, despite the undeniable importance of global pressures, Iranian women themselves are the greatest agents of change. I dedicate this essay to them for their gallant efforts and steadfast commitment in the pursuit of their ideals.

There are obviously a number of issue areas that relate to women's rights, but given the limited scope of this study, I have selected some that show both the persistent hold of traditional values and attitudes, such as in the areas of marriage and divorce, and others where considerable progress has been made over the decades, such as in education and family planning.

THE RIGHT TO LIFE

It goes without saying that the right to life is the most fundamental right, without which there are no other rights. I have chosen to address this issue area to also include freedom from torture and from violence, all of which are addressed in the Universal Islamic Declaration of Human Rights (UIDHR). Changes in this area, however, have been slow in coming and are in a constant state of flux, depending on the political conditions of the moment.

The Islamic Codes enacted in 1982 stipulated the death penalty not only for premeditated murder but also for such "moral offenses" as rape, adultery, sodomy, and habitual drinking of alcohol. These codes specified death by stoning for adultery and even spelled out how the punishment should be carried out.[3] This draconian method of execution involves burying a woman up to the armpits and a man up to the waist and then pelting the victim with stones, neither so big as to kill outright, nor so small as to allow for survival, a method that has been called "an outrage" by Amnesty International and denounced for being "specifically designed to increase the victim's suffering." The punishment had been suspended from 1997, when President Khatami first came to power, until 2001 when it was reintroduced, allegedly as a backlash on the part of conservatives against reforms attempted by President Khatami.[4]

However, in December 2002, when the EU linked trade talks with improvements in Iran's human rights record, Iranian officials assured the EU human rights delegation that stoning was to be abolished in Iran as a form of capital punishment and that henceforth "prison terms and other forms of punishment for the crime of adultery" were to be implemented. Some female parliamentarians have drawn up a bill to eliminate stoning from the penal code. Although it remains to be seen whether the Guardian Council will approve the bill, "the support of leading pro-reform clerics at least offers hope among reformists that the bill will be approved."[5] Meanwhile, the head of the Supreme Administrative Court, Qorbanali Dorri Najafabadi was reported to have said, "The practice {of stoning} has been stopped for a while" and MP Jamileh Kadivar is said to have reported, "To the best of my knowledge, Ayatollah Mahmoud Hashemi Shahrudi has ordered that execution by stoning should be stopped."[6] However, in January 2003 Guardian Council member Ayatollah Gholamreza Rezvani declared there was no substitute for the stoning of adulterers. "Stoning is a sanction for ethical problems such as adultery and there is no other sanction for having intercourse with a married person," he said.[7] And in early February 2003 Iranian state television (IRIB) reported that judiciary chief Hashemi-Shahrudi had said that executions by stoning are seen as a way of "protecting the family" and that "at present, we are not considering a substitute for that punishment." At the same time the English-language version of the official Iranian wire service reported that Judiciary chief Hashemi-Shahrudi had told the EU commissioner that the justice system has decided to replace stoning with another form of punishment.[8] Clearly, the matter is in a state of flux and uncertainty.

While this medieval form of punishment is being debated, many ministries are conducting research on women's human rights. "We study all areas of violence against women, whether in the home or at the hands of the government, and we draw up recommendations," one young woman working for the Ministry of Foreign Affairs proudly said. When asked whether she considered lashing as a violent form of punishment and a human rights violation, she demurred, and spoke of the need to uphold "Islamic" mores and values.[9] The punishment of public lashing for moral offenses continues, although recently its frequency for such violations as "improper veiling" seems to have somewhat diminished.

One area of violence against women that has been notoriously difficult to curb is that of "honor" killings of daughters, sisters, wives. Often mere suspicion of sexual transgressions of even the most innocent kind (such as talking with a boy) are enough to prompt the slaying of a girl or woman in the name of preserving the honor of a man or of his family. Such extreme forms of violence are rare and occur primarily in less developed areas of the country where ancient tribal and feudal norms persist, and where judges and the law have dealt with such murders with some degree of leniency. However, in the year 2000 alone, 24 "honor" killings were reported. Observers believe that many others remain unreported.[10]

Although Iran's constitution prohibits the torture of prisoners, numerous allegations of torture of not only criminals but also political dissidents have been made by families of the detainees and by impartial observers.[11] Addressing the problem, in December 2002 Iran's reformist parliament approved a revised bill banning the torture of prisoners, an earlier version of which had been rejected by the Guardian Council. The new bill bans any form of physical or psychological pressure and specifically outlaws isolation, blind-folding, humiliation, deprivation of food, water or hygienic facilities, nighttime interrogations, and the use of drugs on prisoners. At the same time, EU officials in Tehran for trade talks are insisting on progress on such human rights concerns.[12]

POLITICAL RIGHTS

Another area beset by a continuing tug-of-war between competing values is in the area of political rights. My concept of these rights for the purposes of this essay include due process of law as well as freedom of opinion, expression, and the press.

Although women in Iran have had the right to vote and the right to become members of parliament since 1963, and although Iran currently has 13 women members of parliament, a female vicepresident, a female advisor to the President, and numerous women in various ministries doing research on the status of women, there are certain areas where women had hoped to make greater advances just before and during Khatami's presidency. In his preelection speeches Khatami had extolled women's abilities and even called for "affirmative action" for women because they had been deprived for so long of their rights. "We must act to compensate for that historical void and injustice," he had said, adding, "Iranian women with the right credentials and skills have every possibility of achieving the highest possible positions." Yet, women activists were sorely disappointed when Khatami failed to appoint any women to cabinet positions in his government.

Critics such as parliamentarian Ms. Mosavari-Manesh deplore the seeming inability of Khatami to take significant steps "to provide the means for the participation of women in Iran's social and political arenas."[13] Others claim that even the few advances that have been made amount only to "tokenism."[14] They feel it will take the participation of a large number of women in high political, managerial, and other positions to bring about changes in the current status of women and few expect this to happen any time soon under the current circumstances.

On the one hand, Iran's Civil Law confirms the need for affirmative action on behalf of women because of centuries of privation in the public sphere; on the other hand, it advocates domesticity and veneration of motherhood. Article 3(14) of the Civil Law declares women and men equal before the law; however, it qualifies such equality by making it conditional on being "within the framework of Islamic Law." As a result, wherever women's rights in Islamic Law conflict with the Civil Code, the Islamic Law prevails and women remain legally unequal in many social, personal, and economic areas.[15]

Some observers, however, explain the absence of women in high political office as partially a matter of female cultural attitudes. Many speak of the "need for women to gain self-confidence," so that they will be more willing to seek high office. "It is a matter of values," according to Dr. Nasrin Mosafa, who cites as an example the fact that roughly 30 percent of women college students major in journalism, but that only 1.10 percent of these actually become journalists.[16] Dr. Zahra Rahnavard, President of Al-Zahra College, voices the same opinion. "Women themselves have been reluctant to pursue high political office," she says, citing her own reluctance to run for a seat in the Majlis.

In spite of the seemingly widespread culture of female nonparticipation, however, during the presidential elections of 1997 and 2001 a number of women registered as candidates. Article 115 of the Civil Code states that any capable "*rejal*," meaning a highly competent political and religious figure, can be a candidate for the presidency. However, there has been a heated debate about whether or not the word *rejal* is gender-neutral. Although the Guardian Council disbarred women candidates from running for the presidency, a number of clerics have declared that there is no obstacle to women becoming presidents or supreme leaders (*fuqaha*), so long as they are qualified. So although for now the Guardian Council has interpreted *rejal* as referring only to men, and few realistically expect to see a woman president in Iran any time soon, the debate is ongoing and "the path to the presidency for women is not closed."[17]

Meanwhile, the closing of numerous newspapers, journals, and magazines is adequate testament to the lack of freedom of opinion, expression, and the press. Women writers and journalists have not been spared in the crackdown. Without the due process of law, notable women, such as Shahla Sherkat, Editor of *Zanan* magazine, Mehrangiz Kar, prolific writer and human rights activist, and Shahla Lahiji, publisher of a reformist paper, have been jailed on some pretext or another for their beliefs and writings. Mehrangiz Kar was arrested after her return to Iran following participation in a conference in Berlin where she allegedly was seen without the compulsory Islamic head scarf (*hejab*) while at the conference, for having spoken out against compulsory *hejab* of women in Iran, and "for insulting Islamic sanctities." She was sentenced to four years' imprisonment. Even though she suffered from cancer and had undergone chemotherapy, for months the courts refused to allow her to go abroad for treatment.[18]

She is now living abroad, being treated for cancer. However, her husband, Siamak Pourzand, was arrested in November 2001 and, despite his old age and declining health, was held for several months in solitary confinement without access to his family or lawyer. Charged with espionage and anti-state activities, he was sentenced after a closed trial to eight years in prison, which was raised to 11 years on appeal. As a result of representations on behalf of Mr. Pourzand by EU delegates, he was released from prison on temporary leave, but was hopeful that he would be fully pardoned within months.[19] Unfortunately, however, he was rearrested and remains in prison as of this writing. Clearly, there has not been much improvement in this area.

EDUCATION

There are, however, some areas where there has been dynamic change. One such area has been that of education for women. The UIDHR, Article XXI(a) provides: Every person is entitled to receive education in accordance with his natural capabilities. At the dawn of the revolution, Iran's religious leadership zealously sought to curtail female students' access to what were considered "inappropriate" courses for women. Among these were engineering, veterinary medicine, archeology, and certain technological and arts programs. With the eventual mellowing of zeal, such restrictions were gradually lifted, and slowly but surely educational gains by women became one of the hallmarks of the Revolution. Today more girls graduate from high school than boys and 63 percent of university entrants are women. What is more, progress has been made not only in the major cities and prestigious institutions of higher learning, but also in provincial and rural areas of the country, where women's elementary and secondary education have made enormous strides. In addition, for the first time in Iran's long history, rural parents are allowing their daughters to go to universities and live away from home in order to acquire an education.[20] A recent study points to how rapid has been the change in the acceptability of female higher education. Many women whose fathers did not allow them to continue their education after high school but pressured them into marriage now see their younger sisters allowed the freedom of continuing their studies, postponing marriage, and working outside the home.[21] Without a doubt, education is the single most important engine of change in all human rights issue areas because it is the most effective means of cultural and attitudinal change.

FAMILY PLANNING

Reproductive choice is another area where impressive progress has been made. During the first ten years of the Revolution, the clerical regime advocated a pro-natal policy, closing family planning clinics, banning contraception, encouraging women to stay at home and bear children, and extolling the role of "motherhood" in the Constitution. Ten years later, however, in 1989, the government of Iran, realizing it had a rapidly growing population on its hands for which it could provide neither schooling nor health care, did an abrupt about-face on its former position on family planning. Proclaiming that limiting the number of children a couple had was an Islamic obligation, the government launched a program to control the birth rate, one so successful that it was cited in the 1994 Cairo Conference on Population and Development as one of the best in the developing world. From a high of 4.9 percent population growth, in 10 years time Iran managed to reduce its birth rate to 1.7 percent. This it managed to do without coercion, with the help of UNESCO, and through a massive propaganda campaign, the building of family planning clinics, providing contraceptives to both men and women, and above all, through the education of women and through supporting women's right to choose and control the number of children they have.[22] The latest statistics show a decline of 32 percent in the total fertility rate (that is, the average number of births per woman) between 1996 and 2000, from nearly 3 to 2, which is to say, Iran has achieved replacement-level fertility.[23]

MARRIAGE AND POLYGYNY

In some issue areas, practices are heavily inclined toward traditional cultural norms, with modest movement toward more universal standards. One such area is that of marriage.

Although the UIDHR to which Iran subscribes gives men and women "the right to marry," without any limitations due to "race, color or nationality," unlike the Universal Declaration of Human Rights it makes no mention of "religion." As a consequence, a strong cultural-relativist norm is embedded in the Civil Code of Iran, Article 1059 of which stipulates that while a Muslim man can marry a non-Muslim woman, a Muslim woman cannot marry a non-Muslim man. Nor can she marry a foreign national without "special permission" from the Iranian Government.[24] As a consequence, there are today at least 5,000 Iranian women illegally wedded to Afghan nationals, many of whom have experienced numerous problems because of their status. In December 2002, a bill was introduced to the Majlis to grant Iranian citizenship to foreign spouses of Iranian women. If successfully passed and then approved by the Guardian Council, this bill will regularize the status of thousands of Iranian women married to foreigners, most immediately affecting those who have married Afghan refugees living in Iran, many of whom are now returning home to Afghanistan.[25] Whether that will happen remains to be seen.

The minimum age of marriage has been another area of concern for women's rights. Until very recently, the Civil Code set the minimum age of majority for girls at 9 lunar years, and specified that girls even younger than that could be given in marriage "with the consent of the {male} guardian." This did not mean, of course, that early child marriage was rampant. In fact the average age of marriage in Iran today is 26. Nevertheless, female parliamentarians, outraged that *any* girl could be given in marriage at the age of 9, fought to have a bill introduced to the Majlis to raise the minimum age. Although their initial attempts to raise it to 15 for girls was struck down by the Guardian Council, they finally succeeded in raising it to 13.[26] While still low, this new minimum age is a milestone in that although it is still based on *shari'a* law, it has been modified.

For many women, polygyny continues to be a matter of concern. Although the UIDHR makes no mention of polygyny, Article XIX(a) specifies every person's right "to marry, to found a family and to bring up children in conformity with his religion, traditions and culture." And the Civil Code of Iran adopted after the 1979 revolution allows polygyny for men— up to as many as four wives, and any number of "temporary wives," while a woman is allowed only one husband and may not remarry unless she is divorced. The Civil Code also abrogated the 1975 Family Protection Act that had placed some limits on polygyny.

In 1986, however, the Majlis passed a 12-article law on marriage and divorce that effectively reinstated the Family Protection Act provision that requires permission of a first wife for a husband to take a second wife. Article 12 of the law also grants a wife the right to obtain a divorce if her husband marries another woman without her permission or if, as ascertained by a court, a man does not treat his wives fairly and equally as required by Islamic law.[27]

The 12-article 1986 law also provides for pre-nuptial agreements that are meant to address the criticism the marriage laws have come under for allowing so much latitude for men. It provides for a woman who is getting married to place in her marriage contract a condition granting her a right to divorce if her husband takes a second wife, thus giving her some leverage in preventing polygyny in her marriage. Although it is not clear how many young women are taking advantage of this provision in the law, and although the law is not retroactive and does not help those women who are already married and are faced with having a co-wife, it is one small step in the direction of improvement in women's rights.

Although women activists know they are facing an uphill battle, many of them are tackling head-on the problem of polygyny by undertaking reinterpretations of the Koranic verse that allows multiple marriages for men. Islamist and secular women alike are voicing the notion that the interpretations of the Muslim holy book, the Koran, were written centuries ago when patriarchal, tribal norms held sway. Social conditions today, they say, are totally different and if one looks closely at the verse one sees that the requirement to treat several wives "fairly and equally" is impossible to fulfill. They further cite Verse 129 from Chapter "Al Nesa" of the Koran that states, "you cannot ever treat women equally," negating the possibility of fairness and equity in a polygynous marriage. Although women activists concede that polygyny is not widespread in Iran, they feel it is important to attack the traditions and laws that allow multiple marriage for men and are trying to chip away at an institution which they consider to be the blatantly patriarchal vestige of a bygone era.

Equally controversial is the peculiarly Shi'i institution of *mut'a*, temporary marriage, which allows a man to come to an understanding with a woman, a *sigheh*, to be married temporarily for any length of time acceptable to both of them, whether 10 years or 1 night, upon the payment of a sum of money. At the end of the agreed-upon time, the union ends automatically, unless the couple agrees to renew their understanding. Any children resulting from such a union are legitimate children of the father, who has the obligation to raise them and support them. The mother has no claim on the children or on her temporary husband.[28]

Throughout much of the twentieth century, as Iran modernized, *mut'a* fell into disfavor as being demeaning to women and granting free license for men to indulge their sexual appetites. After the Revolution, however, the institution was revived and encouraged as an answer to what was seen as the promiscuity of "free love" as practiced in the West. Supporters see *mut'a* as beneficial not only in providing "for natural urges," and thus an escape from the repressive sexual strictures of the clerical regime, but also as providing for the legitimacy and maintenance of any offspring resulting from temporary unions. They also consider it beneficial for providing temporary economic security for women, "notably the poor or the thousands of widows from the 1980–88 war with Iraq." Opponents, however, see it as religiously sanctioned prostitution.[29]

DISSOLUTION OF MARRIAGE

Nowhere is the mix of traditional and modern norms and practices clearer than in the dissolution of marriage. Iran's 1979 Civil Code gave men virtually unlimited freedom to repudiate a wife at will, with or without acceptable reason. Women, however, could seek a divorce only if a husband refused to support her, was impotent, or insane, or addicted to drugs, or absent for an extended period of time.

In subsequent years, however, small steps were taken toward modifying the law. In 1986, a law was passed recognizing a divorced woman's rights to a share of the couple's property as well as increased alimony rights. In 1992, after a prolonged struggle among lawmakers, a change was enacted in the divorce laws that allows a woman who was divorced "unjustly and unfairly" to collect payment from her former husband for services rendered during the marriage. However, there appear to be numerous practical impediments to women's collecting such payment, and this law remains largely symbolic.

In 1997, the Civil Code was amended to allow both men and women to go to court to seek divorce in the event of irreconcilable differences, so that the divorce laws were more equitable and it appeared it would be easier for women to obtain a divorce.[30] However, Mehrangiz Kar, a practicing lawyer and prolific writer, observes that, in spite of such reform laws, in practice much depends on the outlook and cultural attitude of the judge of the court.[31] She has seen all too many instances of women trying for years to obtain a divorce, renouncing their *mehrieh*,[32] moving out of the home, giving up any property or alimony claims, and still not obtaining a divorce, because of the traditional, patriarchal attitude of the judge. "Either the law has to be changed to be more specific, or our culture has to change to be less misogynist," she writes.[33]

In August 2002, the Majlis approved a bill that would grant women the right to seek a divorce equal to that of men, setting as grounds for men or women the addiction, mental illness, or violent behavior of a spouse. The Guardian Council rejected the bill as being "contrary to Islamic laws."[34] Two months later, however, it approved the bill to reform sections of the Civil Act, giving women the right to sue for divorce "provided she has a lawyer, or for reasons of hardship."[35] How this plays out in practice remains to be seen.

In battles over child custody resulting from divorce, the laws of Iran long favored paternal custody of children. Until recently, Article 1169 of the Civil

Code provided for a mother to have custody of a son until the age of two and of a daughter until the age of seven. Here, once again, ancient norms and customs held, based on the idea that the father is the sole breadwinner of the family. Transposed to the twenty-first century, that did not make much sense to Iranian women, who more and more work outside the home and contribute financially to the support of the family. In February 2002 parliamentary lawmakers approved a change in the custody law to allow divorced mothers to retain custody of their sons until they are seven. Like other reformist legislation, this bill faced an uphill battle, but was finally approved by the Guardian Council.[36]

Dr. Nasrin Mosafa, a university professor, writer, and researcher for the Ministry of Foreign affairs believes that until Iran's patriarchal culture changes it will continue to be difficult to alter the laws and even when these are reformed they will be implemented in ways detrimental to women. "Through education, through raising of consciousness of the people, we need to change misogynist attitudes toward women. What is more, women need to be more aware of their rights. Once they are aware of their rights they will gain confidence and stand up for themselves more."[37] It is interesting to note that Mosafa attributes women's rights setbacks largely to cultural predicaments. Obviously, it is hard to say whether this perspective is genuine or is a way of avoiding criticism of the regime, of which she is an official.

Many reformist women in high government positions emphasize that the main difficulties Iranian women face are with regard to family law. According to Zahra Shoja'i, Advisor to the President and Head of the Center for Women's Participation, laws as interpreted today are the products of early patriarchal, tribal values of the leading legists who interpreted the Koran and the Islamic tradition (*hadith*) according to the social and economic conditions of the day and according to the prevailing norms of yesteryear. She, like many others, believes that the laws regarding the family should change with the times, that the norms and values of the seventh to thirteenth centuries are not compatible with life in the twenty-first century.[38]

FUTURE PROSPECTS

To state the obvious, the prospects for future improvement in women's rights are inextricably tied to the very future of the Islamic Republic of Iran. No one can predict that future. The rights-of-women issues, like so many other social, political, economic, and cultural issues, are caught in the pangs of a power struggle between the hardline conservative and the reformist political factions. As we have seen, the efforts of both President Khatami and the reformist sixth parliament (*Majlis*) to pass bills to improve women's rights have been repeatedly blocked by the conservative Council of Guardians on the grounds that they are "un-Islamic." The conservative judiciary has time and again posed as the upholder of Islamic morality by claiming proposed reform measures involving women are apt to lead to "loosening morals." And the Supreme Leader Ayatollah Ali Khamenei has called on Iranian women to avoid "feminist inclinations" and to "serve in the posts that do not contradict with their instinctive features." These features, he claims, are "love, peace and kindness."[39]

Beyond the factional sociopolitical struggle, the status of women has suffered also from the adverse side effects of inevitable changes that grip all societies in transition. Urbanization, rural migration to cities, population explosion, and many other such factors have aided the breakdown of family life. Women have not been immune to drug addiction, high divorce rates, inter-generational conflict, growing numbers of street children, and increasing prostitution. Economic ills of the society, including mounting unemployment (6 million out of 70 million by the year 2004), under-employment, and inflation have also taken their toll on women.

Yet, the interaction between internal and global pressures for change also seems to hold the promise of continuous improvement in the life of women. To tick off just a few internal factors, the 63 percent of university entrants who are women now not only outnumber male students in the social sciences, but also are getting close to doing the same in the physical sciences. While the phenomenon is causing much consternation among hardliners, some of whom are calling for a quota system to limit the disproportionate number of female doctors, it is also altering Iran's traditionally patriarchal social makeup. Saeed Leylaz, an analyst, foresees ever greater participation of women in public life. According to him, "In two to three years we will face a wave of qualified women who will be holding mid-management posts."[40]

In addition, young people under the age of 25, who comprise more than two-thirds of the population, did not grow up under the tyranny of the Shah, but under that of the clerical establishment. Through the magic of satellite dishes (which are banned but ubiquitous), through rented foreign videos (also illegal), through foreign books and articles that are regularly translated into Persian, and, above all, through the internet, the level of discourse about women's rights and their consciousness about such rights has risen astronomically. Chafing under the restrictions imposed on them by religious dogma, living in constant fear of being arrested by morals police and subjected to fines or lashing, they incessantly push the envelope to get around the restrictions.

Yet, women are not simply rebelling against social strictures, but also against the tyrannical silencing of dissent, whether manifested through speech, in print, or in dress. Yearning to have a voice in the governing of their country, large numbers of students, increasingly disenchanted with the slow pace of reforms, have demonstrated, at great personal risk, for a civil society, the rule of law, freedom of speech, and freedom from arbitrary detention, imprisonment, and torture. It was largely their vote that brought a reformist president to power, a president who has said: "The world's future belongs to democracy at all levels of governance, advancing ethical, legal and political values based on dialogue and the free exchange of ideas and cultures."[41] But six years after such visionary rhetoric without commensurate actual progress, a new impatient and more confrontational reformist class is emerging, with younger leaders such as Mohammad Reza Khatami, the brother of the president.

Iran's development pattern also holds out hope for the future. A glance at development indicator figures, published by the World Bank, shows that Iran compares favorably with other developing countries, such as China, Brazil, and India, in life expectancy at birth, mortality rate, and adult literacy.[42] What is more, the United Nations Human Development Report for 2002 lists Iran favorably as being among the countries experiencing Medium Human Development in the following categories: gender-related development, priorities in public spending, energy and the environment, refugees and armaments, gender empowerment measures, eliminating inequality in education, eliminating gender inequality in economic activity, women's political participation, status of major international human rights instruments, and status of fundamental labor rights conventions.[43]

Perhaps one of the most consequential domestic factors that could benefit women is the inclusion of a growing number of clerics among the reformers. Although reformist clerics hold little political power and many of them are effectively silenced by the Special Court of the Clergy, they continue to exert pressure on their conservative colleagues. For example, they have set up an Association for Researchers and Teachers to discuss politics once a week and they publish books and pamphlets in which they confront head-on formerly taboo subjects such as the issues of women's rights. Fatemeh Rakei, a female parliamentarian, recently traveled to Qom, the heartbeat and historical citadel of the Shi'a clerics of Iran, to probe the views of several conservative clerics in advance of a bill to authorize Iran to sign on to the Convention for Elimination of all Discrimination against Women. Although both Ayatollah Jannati (Head of the Guardian Council) and Ayatollah Ma'refat, high-ranking clerics, expressed views in favor of such a prospective bill,[44] the Guardian Council struck it down. It will likely be reintroduced at a later date.

Reformist clerics are deeply concerned about the growing disdain for their rank and file among the people, who hold clerics in general responsible for repression by the religious establishment. Reformers among the clerics wonder whether they can ever as a class regain the people's trust, a confidence they enjoyed momentarily in the earliest phase of the revolution. The lifting, in January 2003, of five years of house arrest imposed on Grand Ayatollah Hussein-Ali Montazeri, one of a handful of highest-ranking clerics who had been expected to succeed Ayatollah Khomeini, the father of the revolution, but who had subsequently fallen into disfavor, was greeted by cheers, especially among women. As a dissident cleric who had kept in touch with the people by modern means of communication during his entire incarceration, he promptly resumed his criticism of the excesses of the regime and suppression of freedom of speech and print, despite the constraints of old age and ailing health.[45]

Such dynamics of domestic change are clearly consequential in the drive to better women's rights. But it is worth mentioning that the pressure of the

UN Commission on Human Rights (UNHRC), and such private groups as Amnesty International (AI) and Human Rights Watch (HRW), are not without effect. In 1996, following an unflattering report by the UNHRC, Iran declared that no more visits by the Special Representative would be allowed.[46] In July, 2002 however, the government agreed to re-admit United Nations human rights experts, prompting the high commissioner for human rights, Mary Robinson, to express the hope that Iran's move would lead to long-term cooperation with the UNHRC.[47]

More likely, the influence of the EU, demanding improvement in Iran's human rights record as a condition to successful trade talks, is greater in practice than those of HRW and AI, whose delegates incidentally were denied permission to attend the EU–Iran human rights dialogue in December 2002. Nevertheless, these organizations recognize the importance of such interaction between Iran and the EU for increasing greater awareness of human rights issues in general and with respect to the status of women in particular.

All said, the real agents of change for promoting women's rights are the Iranian women themselves. Most of them are well aware that "it will take a long time." They know all too well that there will be no quick fixes, that cultural attitudes, laws, and their implementation, among other things, will have to change for their fate to change for the better. In interview after interview with Iranian women I was told in Iran, "It will happen, but not in my lifetime." This half-pessimistic, half-optimistic refrain sums up the painful frustration and anxious hope of millions of Iranian women about the future of their rights.

NOTES

1. Normatively, too, the mixture of continuity and change is demonstrated amply in Iran's simultaneous commitment to international legal instruments and Iranian domestic laws. To sample a few of Iran's international commitments to both universalist and cultural relativist documents, Iran was an original participant in the very drafting and signing of the Universal Declaration of Human Rights. Over the years, it has also signed The International Treaty for the Prevention of Trading and Trafficking in Women, The International Covenant of Economic, Social and Cultural Rights, The International Covenant on Civil and Political Rights, and The Supplementary Agreement for Abolition of Slavery and Slave Trade and Activities and Arrangements that are Similar to Slavery. However, in 1990, it approved the Islamic Declaration of Human Rights, which defines human rights according to *Shari'a* and posits: "A woman is equal to a man in human dignity, and she has rights much as she has obligations. She has her civil status and independent financial resources, and the right to keep her name and lineage."

Before the Islamic Revolution Iran's laws were largely based on French and Swiss law in criminal and legal procedures. Even then, however, under the secular rule of the Pahlavi Shahs, family law remained largely governed by the *Shari'a*. After the revolution, although Iran's Civil Law confirmed the need for affirmative action on behalf of women because of centuries of privation in the public sphere, it advocated domesticity and veneration of motherhood. Article 3(14) of the Civil Law declared women and men equal before the law; however, it qualified such equality by making it conditional on being "within the framework of Islamic Law." Both criminal and family law reverted to being governed by the *Shari'a*. Furthermore, the revolutionary leaders amended the constitution to encourage motherhood and domesticity, while at the same time mobilizing women for political, social, and military participation and urging them to exercise their rights of suffrage.

2. Courtesy of the Permanent Mission of the Islamic Republic of Iran to the United Nations.

3. Reza Afshari, *Human Rights in Iran: The Abuse of Cultural Relativism* (Philadelphia, PA: University of Pennsylavania Press, 2001), 42.

4. *The Christian Science Monitor,* July 26, 2001.

5. *New York Times,* January 17, 2003.

6. ABC Online: December 26, 2002.

7. *Hayat-i No,* December 29, 2002. As reported in gulf2000-list@columbia.edu, January 2, 2003.

8. Gulf2000-list@columbia.edu, February 5, 2003.

9. Personal interview, January 2001.

10. *Zanan* 93 (2003), 2–13.

11. Writers such as Reza Afshari, Mehrangiz Kar, and Shirin Ebadi, and reformist newspapers such as *Etemad.*

12. Gulf2000-list@columbia.edu #11. Human Rights and the Gulf, December 15, 2002.

13. FBIS-NES-2002-0829; *Tehran Times* website; August 29, 2002.

14. Mehrangiz Kar, *Mosharekat-e Siyasi-ye Zanan: Mavane' va Emkanat, (Women's participation in Politics: Obstacles and Possibilities)* (Tehran: Roshangaran and Women Studies Publishing, 2000), 41.

15. Ibid., 74–75.

16. Private interview with Dr. Nasrin Mosafa, university professor, writer, researcher for the Ministry of Foreign Affairs, and director of an institute for the study of international relations connected with the School of Law and Political Science of Tehran University, January 2001.

17. Mehrangiz Kar, *Mosharekat-e Siyasi-ye Zanan,* 67–68.

18. *Frankfurter Allgemeine Zeitung,* March 13, 2001; *Agence France-Presse,* June 19, 2001.

19. *Financial Times,* December 18, 2002.

20. Nasrin Mosafa, private interview, January 2001.

21. Mahnaz Kousha, *Voices From Iran: The Changing Lives of Iranian Women* (Syracuse: Syracuse University Press, 2002), 78–144.

22. For a discussion of population control in Iran see Homa Hoodfar, "Bargaining with Fundamentalism: Women and the Politics of Population Control in Iran." *Reproductive Health Matters* 8 (1996); also Nesta Ramazani, "Women in Iran: The Revolutionary Ebb and Flow," *Middle East Journal* 47, no. 4 (summer 1993), 414–16.

23. *Population Today* (Population Reference Bureau, May/June 2002).

24. *Qavanin va Moqararat-e Vijeh-ye Zanan dar Jomhuriyeh Islami-ye Iran, (Special Laws and Regulations for Women in the Islamic Republic of Iran)* (2nd. edn., n.d.) Tehran: Markaz-e Mosharekat-e Zanan. Article 1060 and 17 of the Civil Code: 40–41.

25. IRNA News Agency. Tehran, in English, 1053 GMT December 1, 2002.

26. *New York Times,* August 27, 2002.

27. See note 24 *supra,* 87; see also Nesta Ramazani, " *Women in Iran*": 418; For a discussion of the Koranic verse that requires treating spouses fairly and equitably, see Fazlur Rahman, "Status of Women in the Qur'an in Guity Nashat, ed., *Women and Revolution in Iran* (Boulder, CO: Westview Press, 1983), 45–48.

28. Shahla Haeri, *Law of Desire: Temporary Marriage in Shi'i Iran* (Syracuse, NY: Syracuse University Press, 1989); See also *Zan-e Ruz,* no. 1294 (1369/1990), 55.

29. Robin Wright, *The Last Great Revolution: Turmoil and Transformation in Iran* (New York: Alfred A. Knopf, 2000), 177.

30. See note 24 *supra,* 62–63.

31. Mehrangiz Kar, *Fereshteh-ye Edalat va Pareh-i az Duzakh (An Angel of Justice and a Piece of the Devil)* (Tehran: Roshangaran Press, 1991), 169–73.

32. *Mehrieh,* loosely translatable as "dowry," is a sum of money promised to a bride by a groom, but rarely actually paid at the time of marriage. Rather, a wife usually claims it upon the dissolution of the marriage, when it is supposed to provide the ex-wife with a sum of money. It rarely is actually transacted, and remains an abstraction most of the time.

33. Mehrangiz Kar, see note 31 *supra.*

34. *United Press International;* FBIS-NES-2002-0928, Tehran, IRNA in English 1354 GMT September 28, 2002.

35. Voice of the Islamic Republic of Iran, Tehran, in Persian, December 1, 2002. *BBC Worldwide Monitoring.*

36. United States' Department of State Human Rights Reports on Iran, 2003.

37. Personal interview, January 2001.

38. Personal interview, January 2001.

39. FBIS-NES-2002-0203: February 3, 2002.

40. http://www.iranmania.com/news/ArticleView/Default.asp?NewsC.

41. President Khatami, Special Contribution. *Human Development Report 2002: Deepening democracy in a fragmented world* (New York: Oxford University Press), 64.

42. *Economist,* January 16, 2003.

43. See note 41 *supra,* 149–250.

44. *New York Times,* January 17, 2003.

45. *New York Times,* February 1, 2003.

46. Afshari, *Human Rights in Iran,* 282.

47. *New York Times,* July 26, 2002.

CHAPTER 12

IRAN, DEMOCRACY, AND THE
UNITED STATES

R.K. Ramazani

While the Bush Administration includes Iran in its "Axis of Evil," the Iranian people see this designation as a threat to Iran's historical pro-democracy movement. Decades of mutual vilification between Iran and the United States predated President Bush's moralistic identification of Iran as evil. The hostility between the two countries dates back to the Islamic Revolution of 1979.

The revolution destroyed Mohammad Reza Shah's regime, a longtime strategic surrogate of America in the Middle East, particularly in the oil-rich Persian Gulf. Ayatollah Ruhollah Khomeini, the founder of the Islamic Republic of Iran, endorsed the seizure of the American Embassy and the holding of Americans hostage for 444 days. This single event, perhaps more than any other at the time, fueled American antagonism toward Iran and prompted President Jimmy Carter to break diplomatic relations with Tehran. A quarter of a century later there is still no real indication of a resumption of official relations.

The pro-democracy aspirations of the Iranian people are more than a century old. But never before have so many millions of people struggled so gallantly for democratic values as they are doing today. Young women and men, especially students, are the vanguards of this pro-democracy movement. They are under the age of 30 and form two-thirds of the 65 million or more population of the country. Furthermore, while anti-Americanism is sweeping across the Arab Middle East, the non-Arab Iranians demonstrate genuinely pro-American sentiments.

One objective of this essay is to probe the deeper meaning of the Iranian pro-democracy and pro-American movement. This task is made possible by placing the movement squarely within the broad outlines of the Iranian cultural legacy. The other objective is to show why the people of Iran perceive the Bush Administration's foreign policy as a threat to their pro-democracy movement (1).

THE LEGACY OF TRIPLE POLITICAL CULTURE

Iran's cultural heritage has become threefold in nature over the millennia at least in part because of its remarkable continuity. Despite momentous changes the past has always been present in Iranian culture. Even in comparison with such old civilizations as those of Egypt and Syria "which underwent great changes in the course of two millennia of history" Richard Nelson Frye says, "Iran seems to have preserved much more of its ancient heritage."[1]

Iranian culture took shape between the sixth century B.C. and the seventh century A.D., before the coming of Islam. Pre-Islamic and Islamic cultures both underpin the Iranian sense of identity and one cannot trump the other. Emblematic of these two strands of the cultural heritage is the statement of the Nobel Peace Prize winner Shirin Ebadi. In accepting the prize, she proudly declared to the world on December 10, 2003: "I am an Iranian, A descendant of Cyrus the Great" and "I am a Muslim."

To these two older premodern strands of Iranian cultural legacy must be added a relatively newer one—that of *modernity*. For 200 years, specifically, since the imperial Russian invasion of Iran in 1804, Western civilization has penetrated the Iranian culture. Modern ideas of nation-state, constitution, temporal law, civil society, representative and liberal democratic government, among many others, have been imported by Iranian intellectuals, who in essence believe in the principle of the rule of the people and human rights as opposed to the millennial legacy of autocracy. Modernist Iranians, religious as well as secular, believe that Islam is essentially compatible with fundamental Western political principles. Shirin Ebadi told the sitting of the European Parliament's Committee on Foreign relations in Brussels on February 25, 2004 "Islam accepts democracy. There is no conflict between human rights and Islam." The renowned philosopher Abdul Karim Soroush and intellectually oriented President Mohammad Khatami essentially claim the same, although their conceptions of "democracy" and "Islam" are quite different.

To appreciate the historical longevity of the theory of compatibility of Islam with democracy, let me cite a nineteenth-century example. Mirza Yousef Khan, known as Mostashar Dowleh, argued in his important work *Yek Kalameh* (One Word) that Islam and *liberal democracy* are compatible.[2] He compared the 17-point French Declaration of Rights of Man and of Citizens with the fundamental Islamic tenets and concluded that they correspond. This modern dimension of Iranian culture has also been preserved in history, and in my opinion, just as Iranian rulers cannot exclude either pre-Islamic or Islamic strands from their political culture, they cannot expunge modernity either. They can do so only at the expense of legitimacy of their rule.

Interactions of State and Religion

In all Iran's history no aspect of its triple political culture has escaped the dynamic interactions between religion and state. A thumbnail sketch of this phenomenon in premodern history can reveal the epic significance of the coming of modern ideas, especially democracy, to Iranian society and state.

Religion–state interplay prevailed in Iran during both the Achaemenid and the Sasanid dynasties. The Achaemenids set the foundation of the Iranian state in the sixth century B.C., which lasted for 200 years before the invasion of Iran by Alexander of Macedonia. At the end of Greek and Parthian rules in Iran the Sasanid dynasty revived the Iranian Empire and ruled the state for 400 years before the coming of Islam in the seventh century A.D. Zoroastrian religion interacted with both Iranian dynastic states, but its impact on the politics of the two empires was quite different.

Cyrus the Great, the founder of the Achaemenid dynasty, created the first "world state" and organized the first "international society" in history. In her classic work Adda B. Bozeman writes that in the sixth century B.C. when the tyranny of empires plagued the fabric of community life everywhere, the Persian Empire, vaster than any preceding empire west of China, attained universal peace for some two hundred years, in a large part as the result of tolerant respect for cultural diversity of the subjugated people.[3] Prudence rather than religious ethics of the sixth century B.C., Bozeman says, influenced Cyrus's statecraft.

Richard Frye, the world renowned historian of ancient Iran, tells me that Cyrus's law, or king's law, predated Roman law. It allowed the religious laws of Egyptians, Babylonians, and Hebrews, for example, to stay in force.[4] Herodotus testifies to Cyrus's "statesmanship and liberality"; Xenophon in his *Cyropeadia* finds Cyrus "deserving admiration," above all, for honoring his people "as if they were his own children"; and the Bible reveres Cyrus for liberating the Jews from Babylonian captivity. To date, Iraqi Jews trace their origins to Cyrus's liberation policy.[5]

In contrast, Zoroastrian faith more than state interest sometimes shaped Sasanid politics. In a nutshell, two factors seem to have contributed to this difference. First, at least two of many Sasanid kings claimed some kind of divine qualities. Second, the Magi (*Majus*), or priests, organized themselves into a priestly hierarchy. During the long life of these two earliest Iranian states, king and Magi performed different functions and religion and state remained separate. Frye says that even during the reign of the Sasanids who made Zoroastrianism the official state religion, the head of the priesthood, Mobadan-Mobad, "became the partner of the Shahinshah."[6]

This ancient tradition of separation of religion and state appears to have been cast aside at the turn of the sixteenth century. Shah Isma'il, the founder of the new empire, fused state and religion in 1501. He declared himself to be the "Agent of God," as had a couple of the Sasanid kings.[7] He and his successors, like the Sasanids before them, aspired to universality. While the Sasanids had tried to fuse the state and religion by Zoroastrianizing the state,

Shah Isma'il tried to do so by "Shi'itizing" it. Yet Shah Isma'il's enterprise eventually failed. Roger M. Savory writes that Isma'il failed to prevent the separation of secular and religious powers, which in fact occurred under his successors.[8] Marshall G.S. Hodgson says that by the end of the seventeenth century the religious establishment asserted itself as separate from the state.[9]

While being separate, the clergy, with popular support, gained over time an unprecedented degree of power and authority. In contrast, the state institutions weakened, and the tyranny of kings gradually alienated the people. The intermittent wars between the Safavid and the Ottoman Empires contributed much to state weakness. The Shi'i-Sunni sectarian as well as territorial conflicts at least in part underpinned the outbreak of wars. However, in my opinion, intellectual stagnation and material backwardness were greater causes of state decline.

Western powers had derived their strength from such momentous developments as the Renaissance, Reformation, Enlightenment, and Scientific Revolution. In contrast, fossilized Islamic and pre-Islamic traditions robbed the Iranians of the power of intellectual creativity and material innovation. Historical backwardness exposed state and society to Western imperial encroachments and with it the profound and epic challenge of modernity to the encrustation of millennial traditions.

Two wars with Russia over lost Iranian territories in Transcaucasia and armed confrontation with Britain over Herat in the nineteenth century further weakened Iran. No less consequential, European commercial and imperial expansion sucked the failing state into the whirlwind of European power politics as never before. Rival British and Russian empires penetrated Iranian state and society at least in part by extracting all kinds of commercial and economic concessions from autocratic rulers to such a point that by the end of the nineteenth century Iran was independent only in name. One of these concessions became the first landmark catalyst of Iranian national awakening in modern times. It was granted by Nasir al-Din Shah of the Qajar dynasty to a British company for a tobacco monopoly. Viewed as injurious to both religion and the kingdom, a leading cleric called on people to stop smoking. As a result of the people's protest the Shah was forced to cancel the concession in 1892.

ABORTED DEMOCRATIC EXPERIMENTS

This first popular movement in modern times preceded the first pro-democracy movement in Iran, known as the Constitutional Revolution (1905–11). Throughout Iran's history until this revolution religion had played a leading role in legitimizing the one-man rule of autocratic monarchs. By this time, however, modern-educated Iranians pressed the idea of the rule of the people, rather than traditional tenets of religion, as the fundamental basis for political legitimacy. Enlightened clergy, secular intellectuals, and Bazaar merchants provided the social base of support for the revolution. The tradition of autocracy and monarchical tyranny on the one hand, and unprecedented foreign domination on the other, were the twin engines of the revolution.

Hence the two most fundamental goals of the revolutionary forces crystallized over time into the two enduring principles of independence and freedom in the Iranian political culture. They were enshrined in the Constitution of 1906–07 as inseparable principles. In this light, the Iranian Constitution, like the American Declaration of Independence, symbolized for all time the highest values of Iranian political culture—*independence and liberty.*

By modern standards, of course, the Constitution was not a perfectly liberal democratic document, but against the backdrop of thousands of years of autocracy this essentially democratic first step was remarkable. The very adoption of the Constitution in 1906–07 and the establishment of the first parliament (*Majlis*) in 1906 were unprecedented in Iran's millennial history. Actually, the constitution was modeled after that of Belgium, while at the same time taking into account Iran's own indigenous norms and institutions. The institution of monarchy was retained, although the monarch's power to grant concessions to foreign powers and to conclude international treaties was subject to the approval of the Majlis. For the first time in history people were enfranchised, but the ancient tradition of rigid social stratification limited their freedom to choose their representatives. They could elect their deputies only from among princes, learned clerics (*mujtahids*), nobles, landowners, merchants, and guild members. The representatives of the people in the Majlis could enact secular laws, provided that they were compatible with the criteria of Islamic law (*shari'a*). A committee of learned clerics was to determine such compatibility. But in fact it was never formed.

This first experiment of the Iranian people with democratic government was tragically aborted. The British and Russian empires, although old-time rival powers in Iran, allied themselves in the face of the perceived German threat to their interests in Iran. They partitioned Iran between themselves in 1907. They also forced the departure of the American, Morgan Shuster, whom the Majlis had hired to reform the nation's finances. His classic *The Strangling of Persia* details the machinations of the British and Russian empires to crush the nascent pro-democracy government of Iran. The unholy alliance of the royalist supporters of the Russian puppet Muhammad Ali Shah and diehard conservative clergy, who demanded the establishment of Islamic law as the sole source of law, equally contributed to the demise of the pro-democracy government.

The second opportunity for the Iranian people to try to experiment with constitutional democracy and representative government had to wait nearly four decades. During the first decade Iran nearly disintegrated. Tribal revolts, factional conflicts, a divided government, and the ineffective rule of a young absentee monarch, Ahmad Shah, the last monarch of the Qajar dynasty, left Iranian society in utter chaos. The occupation of the country by warring foreign powers during World War I compounded the disastrous situation. During the 1925–41 period Reza Shah ruled as a secular anti-clerical modernizing dictator like his European counterparts, Hitler and Mussolini. Reza Shah did much to unite the country. He also reduced foreign domination, especially by abolishing the century-old foreign capitulatory rights in 1928. He pushed for modern education, modern economic infrastructure, and a modern professional army. But his renewal of the injurious Anglo-Persian oil concessionary contract has never been forgiven by politically aware Iranians.

The end of Reza Shah's dictatorship left his son Mohammad Reza Shah on the throne in 1941. Being young and inexperienced in politics, he could not immediately continue his father's legacy of dictatorship until after 1953, as will be seen. The Iranians, therefore, call the 1941–53 peiod the era of "Revival of Constitutionalism" (*ehya-ye Mashrutiat*), signifying remembrance of the 1905–11 Constitutional Revolution. The new burst of democratic aspirations was reflected in an unprecedented degree of freedom of expression, press, and party politics. Pro-democracy nationalists, communists, and Islamists battled each other competitively. In the end, the nationalist pro-democracy Dr. Mohammad Musaddiq was popularly elected and reluctantly appointed by the Shah as prime minister.

Musaddiq strongly believed that democratic government could not be established durably unless Iran could free its oil industry, the backbone of its economy, from the clutches of the Anglo-Iran Oil Company. He made the nationalization of the oil industry the first and foremost goal of his government. He depicted for the people his nationalization dispute with the British as a choice between "independence and servitude" (*Esteqlal ya Enqiad*). He argued Iran's case against Britain at the United Nations in terms of moral as well as economic and political aspirations of the people. His preoccupation with the goal of nationalization and ultimately the goal of an "oilless" economy left little room for pushing for democratic reforms. He denied, for example, that the landed aristocracy contributed to the social and political malaise of the nation as much as the control of the Iranian oil industry by the British.

Depicting the choice of his government in such absolute terms as between independence and servitude seems to have left little room for realistic compromise in the dispute with Britain. As laudable as the goal of "complete nationalization" of the oil industry might have been, it is questionable whether Iran's technical and marketing capabilities at the time made such a goal attainable. To date, it is not clear the extent to which his uncompromising conduct reflected his own predisposition or the pressures of various political and religious factions. For example, I was told by Averell Harriman, President Truman's mediator in the Anglo-Iranian dispute, that every time he thought he had reached an agreement with Musaddiq it fell apart. Harriman believed that Ayatollah Kashani, an activist anti-British cleric, had, behind the scenes, kept vetoing Musaddiq's agreements with him.

As will be seen below, a foreign-engineered coup overthrew the Musaddiq government. The coup is often justified in the West in terms of the fear of a communist takeover of his government. In my own research I have found no real evidence that democratically oriented Musaddiq had any pro-communist leanings or that there was any real threat of the capture of his government by the communist Tudeh Party.[10] On the contrary, there is ample evidence to show that the party made every effort to undermine Musaddiq's government, without success. What might have well contributed to his loss of support by some of his staunchest allies was his increasing role as a solo player. He gathered extensive powers to himself even at the expense of the Majlis, the symbol par excellence of Iranian historical democratic aspirations.

On balance, however, foreign intervention more than domestic factional strife, destroyed his government just as in the first Iranian attempt to experiment with democratic politics during the Constitutional Revolution. The CIA, with British help, engineered the coup against Musaddiq's government in 1953, allegedly because of the "contingencies of the Cold War," a euphemism for U.S. competition with the USSR for power and influence in Iran. It took nearly half a century before former Secretary of State Madeleine Albright expressed the regrets of the American government for its intervention against the democratically elected prime minister.

After Musaddiq's fall from power, Mohammad Reza Shah began to flex his dictatorial muscle with full American support. The Weberian concept of Sultanism well applies to the Shah's quarter-of-a-century rule after Musaddiq. His sultanic dictatorship resembled those of Duvalier in Haiti, Trujillo in the Dominican Republic, and Marco in the Philippines. To be sure, his Westernization policies contributed much to the modern material progress of Iran at the start, but backfired subsequently for its superficiality. His expansion of modern education helped the rise of modern intellectual and professional middle classes and increased political awakening among an ever-growing population. But his one-man dictatorial rule left no real room for the development of civil society, freedom of the press, or respect for individual and human rights. His "White Revolution of the Shah and the People," implemented under the pressure of President John F. Kennedy, helped land reforms, but it ultimately worsened the conditions of the peasantry.

His grandiose notion of turning Iran into a "Great Civilization" led to his voracious appetite for buying arms, especially from the United States. He aimed at making Iran one of the five conventional military powers of the world. As an American surrogate, the Shah was anointed as the policeman of the oil-rich Persian Gulf by the Nixon Administration. His popularly opposed status-of-forces agreement with the Pentagon seemed to many Iranians to be a return of the old capitulatory rights of foreign powers. Ayatollah Ruhollah Khomeini called the agreement "a document of the enslavement of Iran." Emblematic of the people's alienation from the Shah and the United States was the revolutionary label applied to the Shah, "the American king."[11]

At a deeper level of analysis, the Shah, together with his father, for half a century trumped the Islamic strand of Iranian political culture. They both tried to legitimize their unpopular regimes by claiming the mantle of the sixth-century B.C. monarch Cyrus. Mohammad Reza Shah in particular put on an absurd show by crowning himself in 1971. He saluted his self-proclaimed ancient predecessor at the ancient king's gravesite at Pasargadae and celebrated the event outlandishly at Persepolis. Cyrus, whom I call a "*secular humanist*," must have turned in his grave over the pretensions of this modern dictator. His secular, but dictatorial, one-man rule was essentially anti-clerical. But to the traditional masses who valued close associations with the clergy at the time, the anti-clerical stance of the two Pahlavi Shahs was viewed as anti-Islamic.

"ISLAMIC DEMOCRACY"

The Islamic Revolution of 1979 ended a quarter of a century of the American economic, political, and cultural domination of Iran, and half a century of the anti-clerical stance of the Pahlavi Shahs. In establishing the Islamic Republic of Iran, the Islamist factions succeeded in seizing power despite the opposition of moderate religious as well as secularly oriented groups, including pro-democracy nationalists. They placed the clergy in control of the state, contrary, many religious and lay scholars believe, to the age-old Shi'i tradition that prohibits the clergy from ruling the state.

But, in my opinion, throughout Iranian history clerical and state functions had been kept separate most of the time. If so, then Khomeini's doctrine of "viceregency of the religious jurisprudent" (*velayat-e faqih*) was unprecedented in Iran's whole political culture, not just the classical Shi'i tradition.[12] He wished to create a "government of God" in Iran, and the Islamic Constitution institutionalized clerical rule, thereby providing for the most extensive powers of the *faqih*.

Yet in retrospect the revolution intensified the time-honored Iranian sense of political independence. The revolution's greatest achievement, in my opinion, has been the emancipation of Iran from direct or indirect control of foreign powers. For the first time in some two centuries Iranian policymakers make their own decisions, realistic or unrealistic, prudent or ideological, and ultimately beneficial or harmful to Iran's values and interests. Iran's deep sense of cultural identity lies behind its fierce commitment to its political independence.

In contrast to independence, the principle of freedom has got short shrift from the very beginning of the revolution. As noted, within the Iranian political culture the principle of independence has been made inseparable from that of freedom. Khomeini himself set the principle of freedom as well as independence in his well-known slogan of "Independence, Freedom and the Islamic Republic," and the Islamic Constitution of 1979 makes these two principles inseparable. Khomeini rejected the age-old secular pro-democracy aspirations of the Iranians. He opposed out of hand the demand of pro-democracy nationalists and Islamic modernists for adding the modifier "Democratic" to the name of the "Islamic Republic of Iran" before the issue was put to a referendum. Ever since 1979 the Islamist factions have opposed democratic competitive politics.

Reformist factions have always argued that the Constitution includes the principle of "the sovereign rights of the nation" and makes the Majlis, the presidency, the local councils, and Assembly of experts elective institutions. In contrast, hard-line conservatives insist that according to the Constitution sovereignty belongs to the *faqih* as the representative of the Hidden *Imam*, and it gives the lion's share of power to him. From the time of the adoption of the Constitution in 1979 to the present time, the irreconcilability of the principle of popular sovereignty and the doctrine of the rule of clergy has been debated.[13]

Just as secular autocracy failed to choke off democratic aspirations, Islamist authoritarianism was unable to stop the gathering momentum of the pro-democracy movement. Ironically, the campaign to Islamicize state and society by fusing religion and state has inoculated an ever-growing population against Islamism and the mixing of state and religion. As noted, about half a millennium ago Shah Isma'il Safavid ultimately failed to prevent the separation of religion and state that occurred subsequently under his successors. Is there a lesson in this significant Shi'i development in history for hard-line conservatives today?

The resounding presidential victory of Mohammad Khatami on May 23, 1997 (*Dovom-e Khordad*) was emblematic of the surge of the pro-democracy movement. Twenty million women and men, 69 percent of the eligible voters, voted for him. By this spectacular action the people in effect tried to throw off the coercive shackles of hard-line conservatives' domination of politics after nearly two decades. Student activists had grown in number over time and a major student organization became active on more than 50 university campuses across the country. But while the reformist camps grew in strength, the diehard conservative factions increased their anti-democratic actions in the name of Islam and the Revolution. Jacques Barzun says:

> Revolutions paradoxically begin by promising freedom and then turn coercive and "puritanical," to save themselves from both discredit and reaction . . . old shackles are thrown off, tossed high in the air, but come down again as moral duty well enforced . . . .[14]

This is exactly what has been going on in Iran. Morals police have been mercilessly enforcing medieval norms of conduct, especially with respect to the veiling of women; the religious establishment has reduced the marriage age and reinforced a husband's unbridled right to divorce, but with some slight backtracking under reformist pressures. Zealous Islamists such as the supporters of the "Party of God" (*Ansar-e Hezbollah*) have broken into private homes, presumably to enforce the norms of Islamic purity.

Political repression far exceeds social dictation based on petrified religious traditions. Rogue members of the previous intelligence service have murdered a number of outspoken intellectuals. Reformist newspapers have been banned and their editors arrested and jailed. Even the reformist representatives of the Majlis have been intimidated and sentenced to imprisonment. The conservative-dominated judiciary has relentlessly suppressed freedom of expression, and sentenced to prison terms enlightened clerics as well as secular-minded reformists. Recently it sentenced to death a university professor (Hashem Aqa Jari) for advocating the need for some kind of "Islamic

Protestantism,"[15] although the final verdict has not yet been handed down. Among all acts of suppression of freedom, the performance of the unelected powerful and conservative-dominated Council of Guardians stands out. The Council has repeatedly disqualified reformist candidates from running in various elections. In the February 2004 parliamentary elections the Council went wild in its arbitrary abuse of power. It disqualified thousands of pro-democracy candidates from participating in elections.

Throughout his terms of office President Khatami has advocated "Islamic Democracy" as the way to accentuate the rights of the people. He has pointed out the urgent need to meet the mounting demands of the people for civil society, the rule of law, government accountability, and freedom of expression and action. He believes that Iran is at "a sensitive juncture of its history."[16] He rejects the fossilized interpretations of Islam, as he criticizes the mimicking of the paraphernalia of Western civilization.

Since further discussion of Khatami's conception of Islamic Democracy is beyond the scope of this essay,[17] I will summarize his statement published in the United Nations *Human development Report 2002: Deepening Democracy in a Fragmented World*. He writes that he accepts the reality of democracy, which has evolved over the past century as a value, rejects any one form of democracy as "the one and final version" and advocates the formulation of democracy in "the context of spirituality and morality."[18] These concepts are, of course, ambiguous. If "spiritually and morality" mean that they should guide social norms, not run the state, it would then seem that his formulation would approach something like the idea of liberal democracy. But since Khatami has so far accepted the doctrine of the rule of the religious leader, his Islamic democracy is still entangled in the contradictory principles of the rule of the *faqih* and sovereignty of the people, which has been debated since the eruption of the revolution in 1979.

In terms of the three stands of Iran's culture, it seems to me that the formulation of "Islamic Democracy" overemphasizes the Islamic strand almost at the expense of the pre-Islamic and modern dimensions of the triple political culture. In contrast, I suggest the concept of "*Persian Democracy*" as a path to durable political order. The two fundamental values of this formulation are justice and liberty. The value of justice is as old as Iranian culture, pre-Islamic as well as Islamic. The value of liberty is not only universal, but has also been embedded in the pro-democracy struggle of the Iranian people for more than a century. This is why in concluding my address to President Khatami on September 4, 2000 at the United Nations I said "I believe there can be no durable political order without justice under the law and no justice without liberty."[19]

What is far more consequential in practice, however, is the decline of the appeal of the Islamic-democracy formula to the people. The disenchantment of reformist factions, especially student activists, with Khatami's failure to fulfill his promised social and political reforms within the context of Islamic democracy, is mounting. The increasing frustration of the supporters of the pro-democracy movement began to show in the second presidential election in 2001 when he gathered 70 percent of the votes, but with a lower turnout as compared with his first election in 1997. Low turnout also showed up in the second local council elections as compared with the first one. It resulted in a dramatic conservative capture of local councils. Above all, the dismal turnout in the February 2004 parliamentary elections showed the deepening apathy and alienation of politically aware Iranians.

These significant setbacks of the reformists no doubt will prolong the grip on levers of power by conservative factions. Their loosening of some restrictions on social behavior in all probability will not have any dampening effects on the growing demand for secular democracy. Once again university students are in the forefront of rejecting Islamic democracy as the solution to the people's grave deprivation of individual freedoms. The debate on the relation between religion and state has never before been as widespread, as sophisticated, and as unremitting as it is today.

This is a profound reality of the Iranian situation today. It is not lost on keen political observers in America. For example, Fareed Zakaria argues "Iran might well hold out the greatest promise for liberal democracy and secular politics in the Middle East. Having lived under Islamic fundamentalist rule, Iranians are now inoculated against its appeal."[20] To cite another example, Thomas Friedman writes: "If the Iranian thinkers and politicians were ever to blend constitutional democracy with refined Islam that limits itself to inspiring social norms, not running a state, it could have a positive impact on the whole Muslim world from Morocco to Indonesia."[21] Can the Bush Administration understand this profound reality of the Iranian pro-democracy and pro-American movement and stop threatening it?

THE BUSH ADMINISTRATION THREATENS THE PRO-DEMOCRACY MOVEMENT

Iranians perceive the Bush administration's threat to their pro-democracy movement against the background of foreign destruction of two constitutional democratic governments in the twentieth century, as explained earlier. The memory of the CIA-engineered coup that overthrew the popularly elected pro-democracy government of Musaddiq in 1953 burns deeply into the collective memory of Iranians and affects the Iranian perception of the American threat today.

Furthermore, this administration's threats are seen in contrast to all previous American administrations' Iran policy since the Iranian Revolution. Despite the confrontational stance of them all, none before the Bush administration had threatened to subvert the regime or attack Iran by military force. The Clinton Administration pursued a policy of "dual containment" against Iran and Iraq. But it never tired of probing the possibility of engaging Iran. This was particularly the case after the election of Khatami to presidency in 1997.

Scholars and officials alike failed to understand the important synergy between Khatami's campaign for social and political reforms and his foreign policy. Essentially, his campaign aimed at democracy at home and peace abroad. It seemed to me at the time that his worldview resembled slightly the democratic peace theory in America.[22] To be sure, after Khomeini President Ali-akbar Hashemi-Rafsanjani had pursued a more pragmatic foreign policy in contrast with the Khomeini era when Iraq attacked Iran. Rafsanjani helped with the release of Western hostages in Lebanon, maintained a policy of strict neutrality in the Persian Gulf War of 1991, and tried to probe openings with the United States. But President Khatami turned on its head Iran's confrontational foreign policy, and earnestly tried to engage America in people-to-people dialogue.

Khatami's conciliatory foreign policy first aimed at relations with the United States, the traditional "Great Satan." In landmark remarks to the American people on January 7, 1998, he drew a parallel between his view that faith and freedom were compatible in the American historical experience as they are in Iran today. He said "the secret of American civilization lies in the Puritan's vision which in addition to worshipping God, was in harmony with republicanism, democracy and freedom." He also complimented the American people by stating that the "American nation was the harbinger of independence struggles on human dignity and rights."

Khatami's reaction as well as that of the Iranian people to the savage terrorist attacks on the Twin Towers and the Pentagon on September 11, 2001 showed that the Iranian conciliatory approach to the Clinton Administration continued early during the Bush Administration. Within hours after the attacks Khatami condemned them as assaults on human dignity and rights. Later he told the United Nations General Assembly: The horrific terrorist attacks of September 11 2001 in the United States were perpetrated by a cult of fanatics who had self-mutilated their ears and tongues, and could only communicate with perceived opponents through carnage and devastation. Even the conservative Supreme Leader, Ayatollah Ali Khamenei, was the first clerical leader in the Muslim world to call for "holy war" (*Jihad*) against terrorism as a "global scourge." Meanwhile, many pro-democracy activists joined 60,000 spectators who observed a minute of silence during a soccer game in Tehran, and many hundreds of young women and men, many weeping, held a candle-light vigil for the American victims of terrorism.

Against this remarkably conciliatory stance of Iran, the Bush Administration's abandonment of the Khatami government and repeated threats shocked both the government and the people of Iran, conservatives and pro-democracy reformists alike. Iranians noticed a drastic reversal of the Clinton Administration's Iran policy, which had probed repeatedly the possibility of engaging Iran in some kind of official dialogue. The record of the recent past reveals that the Iranian perception of the Bush Administration's threat is real.

This is no place to catalogue numerous policy statements that directly or indirectly threaten Iran. The ones that have particularly roiled the Iranians, however, must be mentioned. In his first State of the Union address on January 29, 2002, President Bush vowed to stop the spread of weapons of mass destruction and singled out North Korea, Iraq, and Iran for special mention, calling them an "axis of evil." In his speech of June 2002 to the

West Point graduating class, he said for the first time that the United States would use force preemptively to stop the acquisition of prohibited weapons because America should not wait until threats fully materialize. Washington has repeatedly accused Iran of seeking weapons of mass destruction.

The United States has also included Iran in the category of "rogue states" who sponsor terrorism. The 33-page National Security Strategy document of September 2002 highlighted the threat of weapons of mass destruction that might fall into the hands of rogue states. The implication for Iran was obvious.

President Bush's threat to Iran especially surfaced on two occasions. First, in his second State of the Union speech of January 29, 2003, he repeated the old charges of Iran's search for weapons of mass destruction and support of terrorism. More noticeably, he said, "We also see Iranian citizens risking intimidation and death as they speak out for liberty, human rights, and democracy. Iranians like all people, have a right to choose their own government, and determine their own destiny—and the United States supports their aspirations to live in freedom." The ruse was not lost on the Iranian people, pro-democracy or otherwise. They saw that the president was publicly trying to play the Iranian people against their government. Second, in a controversial statement on July 12, 2002, Bush blasted "the unelected people who are the real rulers of Iran," their "uncompromising, destructive policies" and their "families {who} continue to obstruct reform while reaping unfair benefits." The statement reflected, according to The *Washington Post* of July 23, 2002, the triumph of the hawks in the National Security Council and the Pentagon, and had taken the State department by surprise.

The only official praise for the controversial statement, to my knowledge, came from Zalmay Khalilzad, a senior director of the National Security Council at the time, and the present ambassador to Afghanistan. He, like his hawkish cohorts in the NSC and the Pentagon, is known as one of the "neoconservatives" in the administration. In a statement on August 2, 2002 addressed to the well-known pro-Israeli Near East Policy Institute in Washington, DC, he praised the president's statement for its "specificity and moral clarity." To the Iranians, specificity was abhorrent for its direct attack on their government, and its moral clarity was seen as a simplistic black and white view of Iranian history, culture, and politics, and an astonishing intolerance of complexity. Given the fierce Iranian sense of independence, conservatives and reformists alike got their backs up just as much as they had done in reaction to the "axis of evil" phrase, coined by David Frum.

Powerful neocons with universally acknowledged influence on the Bush Administration have been even more blunt in their threats to Iran. In their well-known book *An End to Evil: How to Win the War on Terror*, David Frum and Richard Perle say that for Iran there can be no reprieve, "The regime must go" because Ayatollah Khamenei has

no more right to control . . . Iran than any other criminal has to seize control of the persons and property of others. It's not always in our power to do something about such criminals, nor is it always in our interest, but when it is in our

> power and interest, we should toss dictators aside with no more compunction
> than a police sharpshooter feels when he downs a hostage-taker.[23]

To cite another example, American Enterprise Institute's Michael Ledeen says besides Iraq "we must also topple terror states in Tehran and Damascus." The Iranian perception of the American threat, however, is not simply shaped by the Bush Administration's major policy statements and the wild commentaries of influential hard-line conservatives who support the administration. The reality of unprecedented American military presence in Afghanistan, Iraq, and Central Asia with thousands of miles of borders with Iran is perceived by Iran as encirclement by an enemy that is the world's sole superpower. Some of the major events that lie behind this perception of encirclement have their roots in Iran's "dangerous neighborhood," specifically, in Iraq and Afghanistan.

The Iraqi invasion of Kuwait and the American-led war against Saddam Hussein's regime in 1991 prompted the United States to increase unprecedentedly its projection of military power into the Persian Gulf region. Until this war the United States had projected its military power from "over the horizon." But immediately after the war it began to increase its military presence in the region exponentially by both unilateral means and bilateral security agreements with the Gulf Arab states.

The first indication of this new policy came in President W.H. George Bush's important address to a joint session of Congress on March 6, 1991.[24] He set the stage then for building a new "framework for peace" and a "New World Order." That order was to replace the Cold War, now that the Soviet Union had disintegrated. The 1991 war provided the opportunity to plant the seeds of Pax Americana in the Persian Gulf. Twelve years later, from Prince Sultan Base in Saudi Arabia, the United States Air Force paraded its slogan of "Global Reach, Global Power."

The September 11, 2001 attacks on New York and Washington seems to have played into the hands of Dick Cheney, Donald Rumsfeld, and Paul Wolfowitz as well as others who had dreamed since the end of the 1991 war of destroying the regime of Saddam Hussein.[25] They and their cohorts managed to capture the mind of Bush with the idea of launching a second war against Iraq to project America's Global Power. A president who had begun his term with an aversion to foreign affairs and nation-building now called for regime change, democratization of the entire Middle East, preemptive war against potential American enemies, and a global war on terrorism. To date, none of the rationales for this second war have seen the light of day. Iraq, we now know, had no hand in 9/11, no ties to al-Qaeda, no weapons of mass destruction, no nuclear program, and no plan to attack the United States.

Just as the Bush Administration's global projection of military power had its genesis in Iran's immediate neighborhood, the seeds of terror attacks on the United States also were planted in the anti-Iranian neighboring state of Afghanistan under the fanatical Taliban regime. To be sure, the terrorists of 9/11 hailed from Saudi Arabia, but their godfather Osama Bin Laden rooted his network in Afghanistan. Iran and the Bush Administration found a common enemy in Taliban-dominated Afghanistan with which Iran had nearly gone to war over the murder of its diplomats. The Northern Alliance, long supported by Iran, helped the prosecution of the American war in Afghanistan.

Iran's help to the Bush Administration in Afghanistan went beyond the prosecution of war. It cooperated earnestly in the discussions in Bonn which eventually led to the establishment of the Afghan Interim Authority under the leadership of Hamid Karzai, not to mention Iran's commitment of more than $500 million aid to the reconstruction of Afghanistan. Yet the administration hawks did everything to undermine any degree of warming of relations with Iran. Zalmay Khalilzad probably played a major part in that effort.

Iran could play a major role in stabilizing both neighboring Iraq and Afghanistan. But as of this writing there is no indication of even limited American engagement of Iran. The devastating earthquake in the historic city of Bam in 2003 and Iran's acceptance of American relief aid appeared momentarily to provide a new opportunity for dialogue between the two countries, but so far they have both demonstrated rare ingenuity in missing opportunities for reconciliation. In public opinion polls the Iranians have shown genuine interest in resumption of relations with the United States, but ever since the Bush Administration began its threats against Iran, the expression of pro-American sentiments has been restrained.

The Bush Administration must borrow a page from the European book if it is genuinely interested in the progress of democracy in Iran. European governments have engaged the Iranian regime in ways to encourage the Iranian people's pro-democracy movement and to improve human rights conditions in Iran. The Europeans apparently understand better than their American counterparts that in the Iranian political culture the quest for democracy has been for more than a century, intertwined with preserving Iran's political independence. They, therefore, refrain assiduously from making threats against Iran. It is ironic that the Bush Administration has failed to grasp the inseparability of the twin principles of independence and liberty in Iran, as they are enshrined in the American Declaration of Independence.

The author of the Declaration and the founder of the University of Virginia, Thomas Jefferson, also believed in promoting democracy as the antidote to autocracy, but by means of liberal education, not the threat of coercion. The following are his memorable words:

"Enlighten the public generally, tyranny and oppression of mind and body will vanish like the evil spirit at the dawn of day."

Notes

I thank W. Scott Harrop of the University of Virginia, Professor Bahman Baktiari of the University of Maine, and Jack Robertson, Foundation Librarian of Jefferson Library for their assistance.

1. See Richard N. Frye, *The Golden Age of Persia: The Arabs in the East* (London: Phoenix Press, 1995), 1–2.

2. I am indebted to Professor Mohammad Tavakoli for providing me with a copy of this work published in Tehran by Nashr-e Tarikh-e Iran, 1912.

3. See Adda B. Bozeman, *Politics & Culture in International History: From the Ancient Near East to the Opening of the Modern Age* (New Brunswick: Transaction Publishers, 1994), 43–56.

4. I am indebted to Professor Richard N. Frye for our exchange of views on ancient Iran. See his classic *The Heritage of Persia: The Pre-Islamic History of one of the World's Great Civilizations* (Cleveland and New York: The World Publishing Company, 1963).

5. See, for example, Josef Wieshofer, *Ancient Persia* (London: I.B. Tauris Publishers, 1996–2001).

6. Personal exchange of ideas with Professor Richard N. Frye.

7. Shah Isma'il also demanded *sijdah*—that is, prostration before him as if he were God. See Rouhollah K. Ramazani, *The Foreign Policy of Iran, 1500–1941: A Developing Nation in World Affairs* (Charlottesville: University Press of Virginia, 1966).

8. See Roger M. Savory, "Religion in the Timurid and Safavid Periods," *The Cambridge History of Iran*, vol. 6, Peter Jackson and Laurence Lockhart eds. (Cambridge: Cambridge University Press, 1986) 610–655.

9. See Marshall G.S. Hodgson, *The Venture of Islam: Conscience and History in a World Civilization: The Gunpowder Empires and Modern Times*, vol. 3 (Chicago and London: The University of Chicago Press, 1974) 16–58.

10. See Rouhollah K. Ramazani, *Iran's Foreign Policy, 1941–73: A Study of Foreign Policy in Modernizing Nations* (Charlottesville: University Press of Virginia, 1975), 231–42.

11. See R.K. Ramazani, *The United States and Iran: The Patterns of Influence* (New York: Praeger Publishers, 1982), 72–124 and 77–78.

12. At this writing I am working on this important theme.

13. See Rouhollah K. Ramazani, "Document: The Constitution of the Islamic Republic of Iran," introductory note, *The Middle East Journal*, 34, no. 2, (spring 1980) 181–204. See also R.K. Ramazani, *Revolutionary Iran: challenge and Response in the Middle East* (Baltimore: Johns Hopkins University Press, 1986, second printing 1988 with an epilogue).

14. See Jacques Barzun, *From Dawn to Decadence: 500 Years of Western Cultural Life* (New York: Harper Collins Publisher, 2000), 35.

15. See the text of Hashem Aqa Jari's talk in "Doktor Sharia'ti va Prozhe-ye Protestantizm-e Islami," *Iran Emrooz*, 1998.

16. *Jam-e Jam*, External TV, Tehran, in Persian 0600 gmt, August 28, 2002.

17. Over the years I have pored over Khatami's work in Persian and English, including numerous interviews and speeches published in Iran and the West. For readers in English I suggest as a start only a few sources here. See Mohammad Khatami, *Islam, Liberty and Development* (Binghamton, NY: The Institute of Global Cultural Studies, 1998); Mohammad Khatami, *Hope and Challenge: The Iranian President Speaks* (Binghamton University: The Institute of Global Cultural Studies, 1997); and Milton P. Buffington, ed., *Meet Mr. Khatami: The Fifth President of the Islamic Republic*, translated by Minoo R. Buffington, Special Study, *Middle East Insight*, Washington, DC (1997).

18. For the text of Khatami's contribution to this report, see page 64.
19. See my "The Role of Iran in the New Millennium: A View From the Outside," *Middle East Policy* VIII (March 2001), 43–47.
20. See Fareed Zakaria, "How to Save the Arab World," *Newsweek*, December 24, 2001.
21. See Thomas Friedman, "Iran and the War of Ideas," The *New York Times,* June 19, 2002.
22. See R.K. Ramazani, "The Shifting Premise of Iran's Foreign Policy: Towards A Democratic Peace?" *The Middle East Journal* 25, no. 2 (Spring 1998), 177–87.
23. Quoted in Patrick J. Buchanan, "No End to War," *The American Conservative* (March 1, 2004).
24. See the text in R.K. Ramazani, *The Future Security in the Persian Gulf: America's Role*, Special Report, *Middle East Insight, Policy Review* no. 2, (1991), 21–25. This monograph foretold that another Persian Gulf war would be visited upon the area in a decade.
25. The debate rages about the real reasons for the Bush Administration's decision to invade Iraq. Did 9/11 cause the war, or did the preoccupation of policymakers such as Dick Cheney, Paul Wolfowitz, and others with toppling Saddam Hussein's regime since the end of the first Gulf war in 1991? This debate does concern the Iranians. They fear that the same policymakers might push for the invasion of Iran after Iraq. I am indebted to Ambassador David Newsom for calling to my attention James Fallows' "Blind Into Baghdad," and Kenneth M. Pollack's "Spies, Lies and Weapons: What Went Wrong," *Atlantic Monthly*, January/February 2004. See also David Ignatius, "A War of Choice and One Who Chose It," The *Washington Post*, November 2, 2003, Sunday, and Mark Hosenhall and others, "Cheney's Long Path to War," *Newsweek* US Edition, November 17, 2003.

CHAPTER 13

CITIZENSHIP AND DEMOCRATIZATION
IN HAITI

Robert Fatton Jr.

INTRODUCTION

The concept of citizenship and its political crystallization are fundamental elements of modern democracy. Democracy, even in its minimal form, is not possible unless its members conceive of themselves as citizens who are inherently endowed with certain rights. The sense of citizenship is thus critical in the process of democratization; its development erodes the legitimacy of authoritarianism and contributes to the rise of civil society.[1] I seek to demonstrate, however, that while the sense of citizenship is an essential prerequisite for democratization, it is not a sufficient condition for the establishment of democracy in poor societies emerging from long years of dictatorship. In such societies, as a close examination of the fall of the Duvalier dictatorship in Haiti shows, citizenship has its limitations because it is easily undermined by the obdurate constraints of scarcity nurturing patron/client relationships and personalistic forms of messianism.

The establishment of democracy faces therefore many obstacles. Material restrictions due to extreme levels of economic underdevelopment generate a *politique du ventre* whereby the conquest of state office becomes *the* prime means of acquiring wealth. Politics becomes a zero sum game: power turns into a risky and indeed deadly conflict in which the losses of some equal the wins of others. In such a situation, power cannot be shared, it is monopolized and those who monopolize it are bent on annihilating those who seek it. By transforming politics into the primary form of capital accumulation, economic backwardness blocks the ascendancy of a bourgeoisie and working class and negates any productive form of capitalism. The absence of real bourgeois and proletarians and the struggles and compromises that their antagonisms entail make democracy fragile and unlikely.

The effects of material scarcity do not constitute, however, the only barrier to democracy. The cultural habits fashioned by the dictatorial legacy nurture authoritarian patterns of thinking and behavior that are difficult to abandon even when the sense of citizenship has begun to implant its seeds. Finally, in spite of fueling the proliferation of rights, the explosion of civil society has no predetermined democratic ending. Reflecting society's inequalities of power, wealth, and income, civil society represents very unevenly the interests of competing and conflicting forces. It sings with the unmistaken voice of privileged and powerful minorities. The democratizing impact of civil society should not be exaggerated; it has clear limitations rooted in structures of profound social inequities that are bound to restrict the exercise of the majority's citizenship.

If Haiti's recent political history is any guide, it is safe to argue that the viability of democracy in poor societies is gravely undermined by at least four major factors: (1) the virtual absence of bourgeois and proletarian actors; (2) the conditions of extreme material scarcity; (3) the persisting legacy of an authoritarian culture; and (4) the vicissitudes of an inequality-ridden civil society. While these factors are autonomous, they simultaneously interact with each other in a dialectical chain of causality. Such dialectical interaction can create certain analytical difficulties; the prevailing empiricism of positivist social sciences is not readily consonant with the notion that autonomy coexists with interaction. As Robert Heilbroner points out, however, the *raison d'être* of the dialectical method is the use of a language that privileges "ambiguities, Januslike meanings, and metaphorical referents...[which] are resistant to examination by the conventional modes of rational thought."[2] In short, democratization in poor societies is full of contradictions and antinomies; it is a process in constant flux with no predetermined outcome.

In the following pages I explore this process as it has been unfolding in the setting of Haiti's tumultuous contemporary history.

THE PREDATORY STATE AND THE RISE OF THE SENSE OF CITIZENSHIP

The Haitian state has historically represented the paradigmatic predatory state.[3] Constituting the "agency of a group or class," the predatory state as Douglas North has argued, functions primarily

> to extract income from the rest of the constituents in the interest of that group or class. [It enforces] a set of property rights that maximize the revenue of the group in power, regardless of its impact on the wealth of the society as a whole... [and] regardless of its effects upon efficiency.[4]

The predatory state is thus a despotic structure of power that preys on its citizens without giving much in return; its total lack of accountability suppresses even the murmurs of democracy. Civil society as well as political society are forced underground; they inhabit what James Scott has called the "infrapolitical"[5] world. This is the "unobtrusive realm of political struggle ... [where no] ... public claims are made, no open symbolic lines are drawn. All political actions take forms that are designed to obscure their intentions or to take cover behind an apparent meaning."[6]

Until the fall of Jean-Claude Duvalier's dictatorship in 1986, the predatory character of the Haitian state inhibited the development of a democratic culture and compelled it into remaining a "hidden transcript." Civil society and subordinate classes were in the "backstage" and expressed only in coded words their outrage at the official justice and authority of those lording it over them. The fall of the dictatorship, however, indicated that subordinates hitherto quiet had finally mustered the means, the resources, and the courage to break their silence. They exploded into the public stage as a collective historical actor in one of "those rare moments of political electricity when, often for the first time in history, the hidden transcript is spoken directly and publicly in the teeth of power."[7]

The marginalized masses—the *moun andeyo*—were finally demanding their humanity.[8] This demand differentiated the uprising against Duvalier from the numerous moments of popular wrath that have characterized Haitian history.[9] The 1804 revolution against slavery and the persistent open and "hidden" forms of popular resistance against the repressive reach of the state indicate clearly that Haitians do not have a particular affinity for, and attachment to, dictatorial rule. Indeed, the old practice of *marronnage*,[10] of exiting first the spaces of slavery and then the regimented arena of a predatory state to create communities of freedom and cooperation, has demonstrated the remarkable capacity of the poor peasant and urban majorities to revolt against, and withstand, the most severe forms of exploitation and domination. The history of *marronnage* is the history of the constant quest for liberty and solidarity by the abused Haitian masses. It is a history, however, that represents more an "exit" from an exploitative society than a determined integration into it on the basis of full equality.

Thus, the revolt of the masses is certainly not a new phenomenon and is indeed a recurring event in the turbulent politics of the island. The conditions leading to, and following the events of 1986 constitute, however, a rupture with the past because the *moun andeyo* were not merely revolting, they were also calling for recognition of their dignity as human beings and for their inclusion in the moral and political fabric of the nation. Having been excluded from the national community, the *moun andeyo* gained their voices; no longer silent, they began to claim their rights to be full citizens. Their struggles for a political, moral, and cultural presence are well encapsulated in the *tout moun se moun*—every human being is a human being— slogan made popular by Jean Bertrand Aristide, the once prophetic priest who eventually became the first freely elected President of Haiti.[11] The idea that every human being was indeed a human being who deserved respect represented a revolutionary *prise de conscience*; it challenged the utterly hierarchical fabric of Haitian society. *Tout moun se moun* symbolized the fact that popular classes had seized upon the idea of citizenship and were bent on establishing "a community of legally equal members"[12] in which they could exercise their "effective moral membership."[13]

Tout moun se moun exploded on the political stage and signaled the vocal awakening of those who had hitherto been excluded from the political, social, and economic game. As Franklin explains:

> By regaining their freedom of political speech and their right to answer, those who were excluded in Haiti regained their humanity and their citizenship; they became historical actors and the subjects of History. Moreover, they broke the monopoly of speech held by "the elite" up to that time and that of power over the right to speak, a monopoly meaning that the "chief's" word, or that of the authorities, could not be gainsaid, thus guaranteeing their monopoly over power. There was a subversion of the relationship of unilateral and unidirectional communication, which was a relationship of power, a relationship of subordination.[14]

Breaking the silence acquired the potency of open revolt and soon became the *Déchoukaj*, the attempted popular uprooting of Duvalierism and Duvalierists.[15] The confrontation between state and civil society, rulers and ruled, was no longer confined to the subordinates' "infrapolitics," it became a very public expression of defiance to those in power. Haitian civil society awakened from the experience of "state terror"; and the "public declaration" of its "hidden transcripts" symbolized its transformation into an effective political force.

The political processes leading to both the emergence of civil society and the gradual articulation of a public discourse speaking truth to power began in Jean Claude Duvalier's hesitant liberalization of the late 1970s. Initially, it was the press that instigated the first wave of public censure of the Duvalierist regime. The radio, in particular, and more specifically Jean Dominique's Radio Haiti Inter, and later on, Radio Soleil, seized the opportunity to break the fear that silenced Haitians during Francois Duvalier's reign of terror. While Jean-Claude Duvalier was rarely if ever targeted for criticisms, the media condemned in increasingly severe terms the authoritarianism of his government. In reality, the media was pushing for a full liberalization that would eventually culminate in a real democracy. The press was liberating *la parole* and reaching a receptive mass audience. Creole that had seldom been used on the airwaves became the hegemonic means of communication and greatly enhanced popular participation. It integrated *le peuple* in the national discourse and gave it the right to be heard and to voice its grievances. Radio stations were not the only thorn in the side of the government; newspapers like *Le Petit Samedi Soir* were also vehicles critical of the regime.

Under pressure from an increasingly vocal media and from American President Jimmy Carter's new policy of human rights, Jean-Claude Duvalier accepted hesitantly the idea of deepening liberalization. He introduced some degree of political pluralism and tolerated the 1979 "election" of Alexandre Lerouge as an independent deputy from Cap Haitien. In addition, in the same year, Grégoire Eugene and Sylvio Claude were allowed to create two non-Duvalierist political parties, respectively the *Parti Démocrate Chrétien*, which would eventually become the *Parti Social Chrétien d'Haiti (PSCH)*, and the *Parti Démocrate Chrétien Haitien (PDCH)*. Both parties condemned the dictatorial system in place, questioned the legitimacy of the presidency-for-life, and called for free elections. Moreover, human rights movements as well as popular organizations like trade unions and students' associations began to resurface and challenge authoritarian rule. The countryside was also mobilizing in the form of peasant cooperatives known as *gwoupmans* to resist repression and rural exploitation.[16]

The government's relaxation of authoritarianism generated autonomous centers of power that began to articulate an alternative future. Instead of strengthening Duvalierism and co-opting opponents, relaxation widened the gap between rulers and ruled and fomented political uncertainties and vacillations. Nongovernmental organizations challenged state corruption and predation and offered material as well as moral and political assistance to a battered population that had received precious little from the oppressive *macoutiste* regime. Civil society was on the verge of outflanking the government; it was opening new possibilities that had hitherto been suppressed. It implanted the seeds of the idea of citizenship and its accompanying demands for political and social rights.

Thus, Jean-Claude Duvalier faced the typical dilemma confronting dictators embarked on liberalization: the process unleashes powerful forces and demands for change that go beyond the constraining limits set by the interests of the dictatorial coalition. While the coalition initiates liberalization to enlarge its base of support, it dreads that a resurgent civil society will undermine its vital interests. As Guillermo O'Donnell emphasizes:

> The first steps of political liberalization usher in the resurrection, the intense repoliticization, of society—a process that soon outpaces liberalization itself… [People] suddenly lose their paralyzing fear of the coercive capacity of the state apparatus. Recently feared figures are now publicly ridiculed. After years of censorship, avid readers find themselves swamped in a flood of publications that…antagonize the existing powers. Various artistic expressions distill long-festering grievances and demands. In other words, civil society, until lately flat, fearful, and "apolitical," reemerges with extraordinary energy.[17]

Thus, while the liberalization of the 1970s contributed to the resurrection of civil society, it opened a Pandora's box with so many uncertainties and dangers for the continued survival of the regime that it felt compelled to resort once again to repressive measures. Ronald Reagan's election to the American presidency in 1980 spelled the end of Jimmy Carter's human rights policy and facilitated Haiti's authoritarian turn. Barely three weeks after Reagan's victory, liberalization came to an abrupt halt on November 28.

On that day, the regime launched a wave of arrests in a brutal effort to decapitate the nascent democratic movement. Over 130 political dissidents, journalists, lawyers, and civil society's leaders were jailed and sent into exile. The press was silenced again and a tense sense of dictatorial normalcy was reestablished. Liberalization, however, had implanted seeds of contestation that would prove impossible to kill.

The Catholic Church soon filled the vacuum left by the removal of civil society's leaders and organizations.[18] It challenged the abuses of *Duvalierisme* and called for social justice and human rights. Regrouping the most active and *engagés* sisters, priests, and religious people, the *Conférence Haïtienne des Religieux (CHR)*[19] condemned immediately the repressive measures of November 28. It called on the government to cease the deportations and arrests of its critics and asserted that the Church "cannot remain silent, for her duty is to make life more humane and people more conscious so all the values of their lives really correspond to true human dignity."[20] The *CHR* espoused the cause of the suffering masses and was part of a growing prophetic movement bent on eradicating "evil" and bringing justice to the poor. At the head of the movement was the radical wing of the Catholic Church, known as *Ti Legliz* (Little Church), which articulated within a Theology of Liberation a devastating public critique of *Macoutisme*.[21] *Macoutisme* came to symbolize everything that was wrong with Haiti: class exploitation, arbitrary political rule, corruption, and state violence. At the Eucharistic and Marial Congress held in Port-au-Prince in December 1982, 120 religious authorities committed themselves and the Church to the struggle against injustice and the "preferential option for the poor." *Légliz sé nou, nou sé Légliz*—The Church is us, we are the Church, they proclaimed.[22] The slogan of the Congress—"something has to change here"—echoed the demands of *Ti Légliz* and the vast majority of Haitians for a massive social, political, and economic transformation.

The prophetic movement found in Pope John Paul II a sympathetic and powerful ally. Visiting Haiti on March 9, 1983, John Paul II echoed the theme of the Eucharistic and Marial Congress and proclaimed, "Things have got to change." With Jean-Claude Duvalier at his side the Pope made a forceful plea for fundamental human rights and dignity:

> Christians have attested to divisions, injustice, excessive inequality, degradation of the quality of life, misery, hunger, fear by many, of peasants unable to live on their own land, crowded conditions, people without work, families cast out and separated in cities, victims of other frustrations. Yet, they are persuaded that the solution is in solidarity. The poor have to regain hope. The Church has a prophetic mission, inseparable from its religious mission, which demands liberty to be accomplished.[23]

John Paul II's homily shocked and angered an embarrassed Jean-Claude Duvalier[24] who had sought a few months earlier to reverse the authoritarian turn taken in November 1980. In April 1982, the president-for-life rekindled hopes of liberalization by calling for the institutionalization of democracy, promising municipal elections, and appealing for a political dialogue with the Diaspora. Duvalier's initiative found, however, few takers; in fact, no sooner had he voiced his democratic intentions than a new wave of repression fell on the country. The leader of the *Parti Démocrate Chrétien Haitien*, Sylvio Claude, and his daughter, Marie-France, were jailed again. Moreover, the Gérard Duclerville affair put an end to whatever reformist illusions may have existed. In December 1982 the police arrested and tortured Duclerville, the director of the Catholic Volunteers, generating unprecedented antigovernmental activities. Civil society and particularly religious organizations openly challenged the dictatorial methods of the Duvalier regime. In a forceful pastoral letter read in all churches, the Conference of Haitian Bishops argued that the difficult period confronting the country "rather than dividing us ... should unify us. Today it is Gerard and those whose names we don't know. Tomorrow, it's us, you, me. Everyone's a victim. Wherever a man is humiliated and tortured, the whole of humanity is."[25] *Radio Soleil* filled the airwaves with a song proclaiming, "The flag of violence has been raised," and for the first time the Duvalier regime was publicly likened to "assassins."[26] The power of the word had finally been liberated and everything was now possible. Neither the release of Duclerville a month before the pope's visit, nor the legalization of certain political parties could halt the revolt of civil society that would ultimately overturn almost three decades of Duvalierist domination.

Duvalier's liberalization was ensnared in contradictions: it promised democratic rule but it never went beyond the constraining parameters of Duvalierism and the presidency-for-life. Initially, it engendered cautious optimism, then increasing frustrations, and finally rage and rebellion. "Liberalization is inherently unstable," as Adam Przeworski has explained:

> What normally happens is ... "the thaw:" a melting of the iceberg of civil society that overflows the dams of the authoritarian regime. Once repression lessens, for whatever reason, the first reaction is an outburst of autonomous organization in the civil society. Student associations, unions, and proto-parties are formed almost overnight.... Thus, on the one hand, autonomous organizations emerge in the civil society; on the other hand, there are no institutions where these organizations can present their views and negotiate their interests. Because of this *decalage* between the autonomous organization of the civil society and the closed character of state institutions, the only place where the newly organized groups can eventually struggle for their values and interests is the streets. Inevitably, the struggle assumes a mass character.[27]

This is the moment when politics becomes "contentious" and the "opportunity structure" widens, providing resources that had been previously inaccessible or inexistent to people who had hitherto been passive. The social movement crystallizes and confronts with sustained popular challenges rulers and office holders.[28] The outcome is generally a change of regime and in a few rare cases a genuine revolution. The point here is that marginalized, and repressed masses organize in unprecedented ways to change their existence. Divisions and vacillations within the ruling bloc offer opportunities for outbreaks of rebellion, which in turn generate further incentives for collective action. What I shall call the structure of obedience and silence falls. In other words, the system of state repression, the patterns of individual and collective behavior, and the moral authority of power no longer hold. The victims of injustice simply stop putting up with their condition; they know now that their sufferings are neither inevitable nor permanent. They have conquered the "sense of inevitability" that had kept them passive for so long; ultimately, they rebuff the routine of obedience.[29] They have become citizens.

In this vein, the Haitian process of democratization reflected civil society's *débordement* of the state. By *débordement* I mean the capacity of civil society to defy and ultimately overwhelm the predatory state and its *projet disciplinaire* through illegal mass political defiance and protest.[30] *Débordement* does not, however, entail a necessary transformation of the state; ruling classes can contain it and limit its impact to a mere change of regime. Moreover, internecine conflicts can fragment the antiauthoritarian coalition into small and opportunistic coteries of self-seeking "big men." In short, civil society's *débordement* of the state is synonymous neither with revolution nor democracy. While it may unleash an uncertain democratization, it is unlikely to generate the profound transformation of the state. Thus, civil society is a potentially liberating factor in any political calculus; but it is not always civil, let alone progressive. It can be quite uncivil; it is replete with antinomies. Embedded in the coercive social discipline of the market, civil society is virtually bound to come to the defense and promotion of private rights and sectional claims.

CIVIL SOCIETY AND THE LIMITS OF CITIZENSHIP

In spite of a strong popular basis, Haitian civil society has remained since Duvalier's fall the preserve of middle and privileged classes whose organizations—rooted in external sources of power and finance—have transformed the country into the *République des ONGs*, "the NGOs' Republic." Rather than constituting a coherent social project, Haitian civil society has tended to embody a disorganized plurality of mutually exclusive projects that are not necessarily democratic. As a plural realm, it broadly comprises three different blocs or spaces articulating respectively the interests of (1) the neo-Duvalierist authoritarian coalition (2) the neoliberal reformist bloc, and (3) the populist *Lavalasian* sectors.[31]

These blocs are not frozen entities, they are internally fragmented, and members of each can move from one to the other. In fact, a strong political opportunism has marked the history of these three blocs, and there is an astonishing circulation of leaders, class fractions, and parties from one sector to another. Dramatic *volte-faces* reflecting very sudden changes of allegiance are common among the political elites and class groupings. Defection and expulsion from "political families" as well as reintegration and cooptation into them are prime characteristics of the conflictive nature of Haitian civil society. For instance, while the "old" ruling class and reactionary segments of the "possessing" class[32] have been the basis of the neo-Duvalierist coalition and supported wholeheartedly the military overthrow of Aristide in 1991, some of their key figures gradually made their peace with Aristide once he returned to power in 1994.[33] Similarly, the "reformists" of the possessing class who had initially backed the *Lavalas* movement, turned their backs on the President and his successor, René Préval, fearing a lapse into a new dictatorship. In addition, the "Lumpen," which had served as the *Macoutiste* foundation of the Duvalier dictatorship and the *attachés* of the Cédras junta, became the *Zinglendos* of criminal bands and the *Chimères* of an increasingly militarized *Lavalas*.[34] Different fractions of social classes had thus taken on different and contradictory positions depending on how distinct historical conjunctures affected their immediate interests. Ideological principles and loyalties were hesitant, ephemeral, and ultimately irrelevant in the political struggle. What mattered was how to be proximate to, and stay in the sites, of state power.

In this respect, both *Lavalas* and the opposition were prisoners of *la politique du ventre*, the politics of the belly, a form of governability based on the acquisition of personal wealth through the conquest of state offices. In a country where destitution is the norm, and private avenues to wealth are rare, politics becomes an entrepreneurial vocation, virtually the sole means of material and social advancement for those not born into wealth and privilege. Controlling the state turns into a zero-sum game, a fight to the death to monopolize the sinecures of political power. The tragedy of Haiti's systemic foundation is that it literally eats the decency and humanity of perfectly honest men and women, transforming them into *grands mangeurs*, big eaters—a rapacious species of office holders who devour public resources for their exclusive private gains. Rather than inviting moral redemption, the immense poverty plaguing the country has generated a generalized pattern of callous indifference and a thoroughly individualistic *sauve qui peut*. Deprivation is so overwhelming that to go on living, Haitians simply pay no attention to it. Like the American journalist Bob Shacochis, the island's rich have always "pushed silently through the shoals of beggars, ignoring their extended hands."[35]

The island's predicament is rooted in this pervasive scarcity. The intensity of the great majority's struggle to escape from its devastating consequences is matched only by the utter determination of most members of the dominant class to avoid at all costs any slippage in their privileged status. These two opposite strategies have contributed to the huge divide separating the dominant class from the masses and have transformed Haiti into "two worlds."[36] The "first world," comprising the wealthy, French-speaking and cosmopolitan minority, displays an utter disdain for the masses and sees politics as a win-or-die proposition. Most mulatto or light-skinned individuals belong to this world. It is antidemocratic and completely excludes the "second world" from its privileges. The "second world" comprises *le peuple* and represents the vast majority of the population. It is overwhelmingly peasant, illiterate, poor, and black. It comprises also the large population living in the squalor of ever-expanding urban slums. The destitution of this second world is overwhelming; according to United Nations' estimates, 65 percent of Haiti's population lives below the poverty line, in rural areas the number rises to an alarming 81 percent. Life expectancy at birth is 51 years for men and 56 years for women, and the mortality rate for infants under the age of five is 125 per thousand.[37] Not surprisingly, Haitians have dubbed the second world as the world consisting of those *san non*, those who "have no name."

It would be wrong, however, to assume that these two worlds exist in mutual isolation. In fact, first and second worlds are relational; they are the opposite sides of the same coin. The second world is defined in relation to the dominant class; it is dependent on it and subject to it. At the same time, the first world's wealth and status derive from its control and taxation of the poor majority.[38] First and second worlds are thus bound together in unequal but interdependent relationships that have generated an enormous gap between the haves and the have-nots. According to conservative estimates, the richest 1 percent of the population monopolizes 46 percent of the national revenue.

The dichotomic structure of Haitian society is also reflected in acute color consciousness.[39] The conflict of color, or the rift between "brown" and "black," exacerbates further social tensions. The line that divides Haitians most, however, is not color but class. The line of class expresses the fact that while first and second worlds are thoroughly intertwined in an exploitative material web, the dominant class has always sought a total moral and psychological dissociation from *le peuple* whom it has dehumanized into meek and servile creatures. Armed with an acute sense of cultural superiority, the dominant class is conscious of its privileged status in the social order, and bent on defending it. It has an adversary position toward subordinate classes, whom it regards with scorn and fear. By dissociating itself from *le peuple* through its language, religion, education, and etiquette, the dominant class has tried to validate its elevated position and its claims to a natural right of governance. As for the poor, the vicissitudes of eking out a meager existence have rarely afforded them with the opportunity of changing their miserable circumstances. Historically, emancipation has assumed the form of personal gain or fulfillment to the detriment of any collective purposeful action.

Haiti provides a paradigmatic case of the difficulties—if not the impossibility—of establishing democratic rule in extremely poor nations plagued by a despotic inheritance. By generally reflecting the lopsided balance of class, racial, and sexual power, the agencies of civil society inevitably privileged the privileged and marginalized the marginalized. Civil society's plurality does not entail an automatic and equal representation of the whole polity. Civil society is not an all-encompassing movement of popular empowerment and economic change. It is simply not a democratic *deus ex machina* equalizing life-chances and opportunities; crippled by material limitations and class impairments it should not be confused with a "civic community." The latter, as Robert Putnam defines it, is

> marked by an active, public-spirited citizenry, by egalitarian political relations, by a social fabric of trust and cooperation . . . Citizens in a civic community, though not selfless saints, regard the public domain as more than a battle-ground for pursuing personal interest.[40]

It is clear, however, that the seeds of the civic community cannot be planted without a dense civil society regrouping a vast network of associational life. Moreover, if these seeds are to flourish in Haiti they must privilege the *Lavalasian* civil society—the popular civil society of the subordinates. Born of the difficult struggles against the Duvalier dictatorship, and expressing a sense of communal defense against the abuses of state power, popular civil society has certain social-democratic impulses, which are almost always expressed in a contradictory and demagogic populism.

Upon seizing power, *Lavalasian* civil society called for the establishment of three different types of citizenship resembling closely those identified by T.H. Marshall more than 40 years ago.[41] First, the movement demanded political citizenship—"the right to participate in the exercise of political power." Second, it sought civil citizenship—"the rights necessary for individual freedom." And finally, it claimed social citizenship—"the right to a modicum of economic welfare and security [and the] right to share to the full in the social heritage and to live the life of a civilized being according to the standards prevailing in society."[42]

Moreover, *Lavalas* injected into Marshall's idea of citizenship a populist vision of participatory democracy transcending mere parliamentarism and privileging initiatives "from below." It is true that in practice this meant often a form of "mob rule" that intimidated not only enemies, but also potential allies. At its best, however, it aspired to suppress the obdurate class impairments of Haitian society. It was an attempt, albeit clumsy and demagogic, to articulate the democratic collective rights of traditionally marginalized classes. In fact, historically, these classes have been the most forceful promoters, defenders, and supporters of democracy, not because of some special vocation or superior moral qualities; but rather, because they have a vested interest in democratic rule. History has taught them the simple lesson that without such rule, they will be consistently excluded from political participation; in the absence of democracy, they will be condemned to suffer the most acute moral indignities and material deprivations that such exclusion entails.

Haiti's democratic transition would have never materialized without the protests, the strikes, and the energy of the *Lavalasian* movement. The movement represented the determinant social force that compelled predatory rulers into accepting pacts to form a politically more accountable regime and ultimately to organize free elections. In addition, the old practice of *kombit*—a Creole word meaning "working together"—which inspired cooperative work and transcended the pursuit of self-interest, resurfaced with the rise of the new "democratic movements." These movements originating from the grassroots and comprising trade unions, peasant associations, religious groups, and diverse professional organizations are the very foundation on which democratization can plant its seeds and flourish. This is not to espouse an easy triumphalism. On the contrary, the democratic movements face huge difficulties; they lack resources and organization and many of their leaders have become fatigued and demoralized by years of struggles and apparent failures, and some have even degenerated into opportunistic *grands mangeurs*.[43] To this extent, members and cadres of the movements can easily lapse into irresponsible, unaccountable, and self-seeking behavior. They can become the new patrons as they use their position to acquire illicit wealth and power, and followers can revert to being clients seeking modest prebends to survive in an environment of utter scarcity.

Moreover, the subordinate classes comprising *Lavalas* are capable of thoroughly undemocratic and indeed despotic, cruel behavior; the incidence of violence by and against the "lumpen-poor," and the practice of necklacing show that they do engage in atrocities and can be mobilized for the worst kind of brutality. It is also true that their poverty invites unpredictability; the highest bidder can buy them for noble as well as despicable acts. After all, for mere pittances, *Macoutes* perpetrated savageries against their own brothers and sisters of the subordinate classes.[44] As Amy Wilentz put it:

> [The] lumpen are traditionally fickle. At moments of great historical change they may support you for your ideas, for your words. But many among them can be bought. In times of plenty they are loyal, but when was the last time Haiti had experienced a time of plenty? And in times of penury their support can be and often is purchased by the highest bidder—and for very little. For a dollar they'll demonstrate. For twenty, may be less, they'll torture, they'll burn, they'll kill, they'll assassinate.[45]

Thus, the participation of subordinate classes in savage moments of cruelty indicates clearly that they have no spontaneous or necessary democratic vocation. They are more interested in democracy than any other class, because they have more to gain from it, but they can hardly claim a higher morality. They simply do not have the monopoly on virtue.

Moreover, subordinate classes carry the burden of their political *habitus*[46]— the system of "dispositions acquired through experience" that shapes particular types of behavior at particular historical moments—they believe in the emancipatory powers of a providential, messianic leader. It is a belief that has undermined political organization and parties. It facilitates the rise of "patrons" and demagogues and obliterates the need for institutional structures of governability. "One manism" becomes the dangerous norm. It is not surprising, therefore, that given his humble roots, fiery homilies against Duvalierism, and brave struggle against injustice, Aristide came to symbolize the prophet whom God entrusted with leading his people to rise against their oppressors. Bypassing forms of collective accountability and decision-making, Aristide condensed unto himself whatever transformative project the *Lavalasian* bloc may have had.

Faced with the prospects of defeat in the presidential elections of 1990, the bloc abandoned the long-term strategy of building a coherent mass-popular party, dumped its lackluster candidate, and opted for the charismatic anti-Duvalierism personified by Aristide. While this choice ensured victory at the polls, *Lavalas* became so closely identified with Aristide that it lost its autonomy. By idealizing him as the *sauveur*, the movement nurtured unintentionally the cult of his personality and ultimately imparted to him supernatural powers. Aristide's courageous struggle against Duvalierism and his ability to survive numerous attempts on his life gave him an aura of mystical invincibility and a unique popular legitimacy.[47] In addition, Aristide was a man of the people, he was born into the peasantry, he knew the vicissitudes and hardships of poverty and his ascension to the clergy transformed him into the voice of the poor. He came to symbolize the fight against *Macoutisme* and the embodiment of popular aspirations. The main slogan of his presidential campaign unambiguously stated: *Titid ak nou, nou se Lavalas*—Together with Titid, together we are *Lavalas*. Aristide was the prophet as well as the prince. He was indeed *Lavalas*.[48]

This personification of *Lavalas* had its dangers; it contained the seeds of a possible new *derive totalitaire*. Aristide could easily assume a presidential monarchism bent on suppressing any alternative, independent power. As Jean and Maesschalck have argued:

> Even if "Lavalas" is a popular phenomenon, the "source" of the tendency resides in a personal initiative founded on "charismatico-religious" leadership. Despite its complexity and contradictions, the entire movement is built on one person, the only one capable of giving it impetus and possibly a new start. There is thus a relationship of complete dependence with respect to the leader. But the latter engages in anti-organizational practices that have already been recognized: among his entourage, no group has been able to assume an autonomous status and to develop freely. This leader spontaneously opposes any form of control over his power. His action is stimulating for individuals and can produce mass effects, but it has a destructuring effect on a group striving to form and to organize on an objective basis . . .
>
>   This religious relationship with the leader has a wait-and-see attitude with relationships based on confidence and any questioning of leaders, autonomy of those acting, and critical conscience being perceived as treachery. The priest-president has neutralized the development of enlightened political judgment among the masses. He has transformed a demanding people into the well-behaved masses trusting in the occurrence of miracles.[49]

Thus, the personalization of power has gravely undermined the coherence, independence, and organizational drive of the *Lavalas* movement. Paradoxically, the emergence of Aristide as the messianic *chef* has emasculated the transformative potential of subordinate classes who are now a disorganized mass awaiting personal salvation. This creates a serious danger because a meaningful democracy is impossible without strong institutions that have precedence over any individual leader. In fact, in Haiti as elsewhere, democratic accountability depends on political parties and other associations privileging the mobilization, and organization of subordinate classes. Clearly, accountability can be established neither on the basis of these classes' occasional public eruption at election time, nor on their exclusion from the sites of power.

Subordinate classes must be the prime agent of any democratic alternative; they are, by far, the largest and most exploited social group. It is true that class does not exhaust the multiplicity of possible forms of individual identity, and that it does not constitute the only exploited collective agent. In Haiti, race and gender have also historically represented identities of oppressed categories. Class represents, however, the only agency capable of uniting the disparate interests of the marginalized. Moreover, subordinate classes, in spite of their uncertainties, remain the class most objectively committed to the realization of democracy because of their productive relations and life experiences.[50] Of all social actors, as Rueschemeyer, Huber Stephens, and Stephens [51] have argued, it is the subordinate classes, and in particular the working class, that have the most interest in the expansion, and preservation of democratic practice.

The problem, however, is that in a poor and nonindustrialized country like Haiti, the working class is virtually nonexistent. Representing only 9 percent of the labor force, the working class numbers less than 200,000 people out of a total population of over 7 million inhabitants.[52] Mostly concentrated in Port-au-Prince, the working class is organized in three main unions: the *Centrale Autonome des Travailleurs Haitiens (CATH), the Federation des Ouvriers Syndiques (FOS), and the Confederation des Travailleurs Haitiens (CTH). CATH, FOS,* and *CTH* claim to have respectively 176,000, 200,000, and 3,000 members; the vast majority of whom, however, are peasants.[53] While the working class knows that it needs democracy to survive and improve its conditions, its small size and ideological divisions have significantly limited its impact on the process of democratization. In turn, these limitations have contributed to both debilitate democratization itself, and generate the conditions for predatory democracy. To put it bluntly, democracy is impossible without the empowerment of subordinate classes and in particular the working class.

For, if Barrington Moore's famous *bon mot*—"No bourgeois, no democracy"—is correct, it necessarily implies its antinomy: "no working class, no democracy."[54] A classical bourgeoisie cannot stand in mid-air, its successful existence presupposes a working class from which it extracts the economic surplus required for capitalist economic activities. Thus, the debate about whether it is the working class or the bourgeoisie which is responsible for democratic rule is artificial; the two classes are opposite faces of the same coin. Their conflicts and struggles are the very stuff generating the historical compromises from which ultimately results democracy.[55] As Rueschemeyer, Huber Stephens, and Stephens have argued:

> The chances of democracy . . . must be seen as fundamentally shaped by the balance of class power. It is the struggle between the dominant and subordinate classes over the right to rule that—more than any other factor—puts democracy on the historical agenda and decides its prospects. Capitalist development affects the chances of democracy primarily because it transforms the class structure and changes the balance of power between classes.[56]

In Haiti, however, such balance of power has not crystallized because capitalist development has produced a bourgeoisie and a working class that are both at best utterly small, embryonic, and fragile. Their unreservedly precarious existence explains in part the very weak character of the country's democracy. Indeed, democracy in its minimal liberal form is the result of a balance of forces between contending classes—the bourgeoisie and the working class—and when these classes are absent, democracy is at best hesitant and indeed predatory. Predatory democracy is a regime where very imperfect trappings of liberal democracy coexist with the Hobbesian struggle to monopolize the few sites of public power with access to wealth and privilege. It is true that constitutional constraints on executive authority exist and that the press remains free, but this struggle for office takes very brutal forms of intimidation and is ultimately determined by controversial and often fraudulent elections. Predatory democracy is thus a hybrid of authoritarianism and polyarchy. In this sense, while it represents a more accountable political system than the typical dictatorship, it suffers from the excesses of presidential monarchism and the intense competition for the appropriation of prebendary gains, both fundamental elements of despotic rule.

Moreover, the very secondary role of the tiny working class in the struggle for democracy has meant that the populist and opportunist tendencies of the Lavalasian petite bourgeoisie have had free rein. In their quest for political supremacy, different factions of the petite bourgeoisie took charge of the democratizing process and marginalized the working class as well as other forces "from below." The subordinate bloc gradually lost its autonomous capacity to pressure power holders. Instead, the burdens of extreme scarcity prevented it from sustaining its efforts and defending its own interests. Exhausted, demoralized, and indeed starving, the bloc fragmented and its strength dissipated. Whatever was left of the bloc became an appendage of other social forces. As Jean and Maesschalck have explained:

> Destabilized at the most profound level of their being by misery, confronted daily with their despairing situation, many militants are unable to invest in long-term construction. Even if they show a desire to do so, they have neither the strength nor the means. From this stems the tendency to lapse into *sterile activism*, to put their efforts into acts that are as dangerous as they are spectacular and with an impact that rarely extends beyond the short term and the ephemeral.[57]

The Haitian case indicates that popular pressures are decisive in the crystallization of the moment of democratization. Unlike the long-term processes of implementation and consolidation, the moment of democratization, however "electrical" and dramatic, is an ephemeral historical event that promises much but offers little by itself. It signals the exit of the dictator and his most immediate cronies, and the beginning of a complicated period of political uncertainties. By compelling the dominant classes and middle sectors to opt for democratization, popular pressures and mass mobilizations are the determining force engendering the crisis that provokes the fall of authoritarianism. In the aftermath of the democratizing moment, however, these popular forces gradually lose steam and are eventually co-opted by better-organized and financially more independent social forces.

Formerly excluded parties and *chefs* of the dominant classes and middle sectors begin to re-impose their power, a task facilitated by the material scarcity and political exhaustion facing subordinate groups. Indeed, these emerging *chefs* can literally buy popular support to advance their own strategic interests. Such support derives from the opportunistic response of poor people for access to basic resources such as food, employment, and health; it has little to do with political loyalty.

The relative decline of popular forces weakens the democratic impulses of democratization, which becomes the affair of the dominant classes and the petite bourgeoisie. The tensions and conflicts between dominant classes and working classes that have traditionally resulted in the development and consolidation of liberal democracy have now been displaced by struggles between ascendant and falling groups of the dominant classes and middle-sectors. The outcome is the predatory type rather than the liberal form of democracy rooted in the classical struggle between bourgeoisie and proletariat.[58]

Haiti's predatory democracy is thus the result of two fundamental factors embedded on the one hand in the persisting legacy of a dictatorial *habitus*, and on the other in the fragility and indeed virtual absence of both a productive bourgeoisie and a large working class. These two factors are mutually reinforcing and have tended to generate a perverse dictatorial cycle from which Haitian society has yet to extricate itself. The utter underdevelopment of the bourgeoisie and working class has contributed to the profound backwardness of the economy, which in turn has nurtured *la politique du ventre* and its accompanying authoritarian propensities. This has meant that subordinate classes are overwhelmed by the daily struggles of eking a miserable existence. They have little time or energy to mobilize as a collective agent bent on democratizing a system that has historically condemned them to a life of squalor. In the rare historical moments when they do, such as in 1986 and 1991, they soon find themselves exhausted, marginalized, and co-opted into patterns of individualized patronage and clientelism. *La politique du ventre* ultimately compelled them to confine their newly acquired sense of citizenship to the "backstage." To paraphrase Brecht, while the old is dying, the new is struggling to be born.

NOTES

1. As Doug McAdam, Sidney Tarrow, and Charles Tilly, contend in *Dynamics of Contention* (Cambridge University Press, 2001), 266:

   Citizenship consists . . . of mutual rights and obligations binding governmental agents to whole categories of people who are subject to the government's authority, those categories being defined chiefly or exclusively by general relations to a specific government rather than by reference to particular connections with rulers or to membership in categories based on imputed durable traits such as race, ethnicity, gender, or religion. Citizenship fortifies breadth and equality of political participation as it defines boundaries between segments of the population that are and are not eligible for different degrees of binding consultation and protection. Democratization means any net shift toward citizenship, breadth of citizenship, equality of citizenship, binding consultation, and protection.

2. Robert Heilbroner, "The Dialectical Approach to Philosophy," in Tibor R. Machan ed., *The Main Debate: Communism Versus Capitalism*, 16.

3. See Robert Fatton, Jr., *Haiti's Predatory Republic: The Unending Transition to Democracy* (Boulder: Lynne Rienner, 2002); Alex Dupuy, *Haiti in the World Economy* (Boulder: Westview Press, 1989); Michel-Rolph Trouillot, *Haiti: State Against Nation* (New York: Monthly Review Press, 1990); and "Haiti's Nightmare and the Lessons of History," *NACLA*, vol. XXVII, no. 4 (Jan/Feb 1994): 46–51.

4. Douglass North, *Structure and Change in Economic History* (New York: Norton & Company, 1981), 22–28; see also: Peter Evans, *Embedded Autonomy* (Princeton: Princeton University Press, 1995), 12; Robert Fatton, Jr., *Predatory Rule* (Boulder: Lynne Rienner Publishers, 1992).

5. James C. Scott, *Domination and the Arts of Resistance* (New Haven: Yale University Press, 1990).

6. Ibid., 183, 199, 200.

7. Ibid., p. xiii.

8. The *moun andeyo* is the Creole word for the "outsiders," those who are not part of the nation and are excluded from its benefits and recognition. See Gérard Barthélémy, *Le Pays en Dehors* (Port-au-Prince: Editions Henri Deschamps, 1989).

9. Alain Turnier documents well these "moments of madness" and retribution that have marked Haiti's history since it became an independent nation. See Alain, *Quand la Nation Demande des Comptes*, 2ème ed. rev. et corr. (Port-au-Prince: Editions Le Natal, 1990).

10. Jean Fouchard, *Les Marrons de la Liberté* (Port-au-Prince: Editions Henry Deschamps, 1988); Carolyn E. Fyck, *The Making of Haiti* (Knoxville: The University of Tennessee Press, 1990); Robert Maguire ("Bootstrap Politics: Elections and Haiti's New Public Officials," *The Haiti Papers*, The Hopkins-Georgetown Haiti Project, no. 2, February 1996, 3) describes *marronnage*—in Creole *mawonaj*—as "a strategy from the country's past and evocative of runaway slaves that boils down to resistance through elusiveness." When threatened, leaders and groups blended into the woodwork until it was safe to re-emerge. This practice would serve them well, not just during their early days of organization, but in the future. Using *mawonaj*, Haiti's evolving

grassroots movement survived Duvalier and the rapacious military regimes led by Henri Namphy, Prosper Avril, and finally, Raoul Cedras and his cohorts."

11. Jean Bertrand Aristide, *Tout Moun Se Moun, Tout Homme est un Homme*, (Paris: Seuil, 1992).

12. Anthony Smith, "The Origins of Nations," in John Hutchinson and Anthony D. Smith, eds. *Nationalism* (Oxford: Oxford University Press, 1994), 153.

13. Ernest Gellner, "Nationalism and Modernization," in John Hutchinson and Anthony D. Smith, eds., *Nationalism* (Oxford: Oxford University Press, 1994), 55.

14. Franklin Midy, "Changement et Transition," in Gerard Barthelemy and Christian Girault, eds., *La Republique Haitienne* (Paris: Karthala, 1993), 206. Translated from the original French quotation:

> En recouvrant la faculté de la parole politique et le droit de réplique, les exclus d'Haiti ont recouvré leur humanité et leur citoyenneté, ils sont devenus acteurs historiques, sujets de l'Histoire. De plus, ils ont cassé le monopole de la parole, jusque-la detenu par "l'élite," et celui du pouvoir sur le pouvoir de la parole, monopole qui faisait que la parole du "chef" ou des "autorites" était sans réplique possible, donc qui garantissait leur monopole du pouvoir. Il y a eu subversion du rapport de communication unilatérale et unidirectionnelle qui était un rapport de pouvoir, un rapport d'assujettissement.

15. *Déchoukaj* is the Creole word for "uprooting"; it became the slogan of the popular Lavalassian forces in their attempt to eradicate Duvalierism and punish Duvalierists.

16. Among these, the largest and oldest, The Peasant Movement of Papaye (MPP), founded in the early 1970s, remained clandestine until the fall of Jean-Claude Duvalier in 1986. Since then it has become one of the most important popular organizations in the fight for democracy.

17. Guillermo O'Donnell, *Counterpoints* (Notre Dame: University of Notre Dame Press, 1999), 122.

18. See Anne Greene, *The Catholic Church in Haiti*, (East Lansing, Michigan State University Press, 1993), 129–209; see also: Ferguson, *Papa Doc, Baby Doc. Haiti and the Duvaliers*, 75–77 (New York: Basil Blackwell, 1987).

19. Anne Greene describes the CHR as "the collective conscience of the Church during the Jean-Claude presidency. The preponderantly female organization (over 80 percent of its 16,000 members are women), repeatedly took issue with the government against injustices and encouraged and pressured the [Conference of Haitian Bishops] to take a stand" (ibid., 141).

20. Ibid., 134.

21. See Jean-Bertrand Aristide, *In the Parish of the Poor*, (New York: Orbis Books, 1991). The Creole word *Tonton Macoute* means "the bogeyman" in Haitian popular folktales; the *Tonton Macoute* was the name given to the brutal paramilitary force created by Duvalier. *Macoutisme* refers to the system of terror imposed by François Duvalier. For a fuller discussion of the term, see Gérard Barthélemy, *Les Duvalieristes Après Duvalier* (Paris: Editions l'Harmattan, 1992), 44–46; see also, Laguerre, *The Military and Society in Haiti* (Knoxville: University of Tennessee Press, 1993) 114–17; and Trouillot, *Haiti: State Against Nation*, 152–56.

22. Greene, *The Catholic Church in Haiti*, 134–35.

23. Quoted as cited in ibid., 138.
24. Ibid. See also, Moise and Ollivier, *Repenser Haiti*, 79–80; and Ferguson, *Papa Doc, Baby Doc. Haiti and the Duvaliers*, 75–77 (New York: Basil Blackwell, 1987)
25. Quoted as cited in Greene, *The Catholic Church in Haiti*, 144.
26. Ibid.
27. Adam Przeworski, *Democracy and the Market* (Cambridge: Cambridge University Press, 1991), 58–59.
28. Sidney Tarrow, *Power in Movement*, 2nd ed. (Cambridge: Cambridge University Press, 1998).
29. Barrington Moore, Jr. *Injustice: The Social Bases of Obedience and Revolt*, (White Plains: M.E. Sharpe, 1978).
30. Fatton, *Predatory Rule*, 105–06.
31. The program of *Lavalas* is contained in: Opération Lavalas (OL), *La Chance qui Passe* (Port-au-Prince: Opération Lavalas, 1990); and *La Chance à Prendre* (Port-au-Prince: Opération Lavalas, 1990). See also Alex Dupuy, "The Prophet Armed: Jean-Bertrand Aristide's Liberation-Theology and Politics," unpublished manuscript, 1995.
32. I see the Haitian dominant class as comprising two broad factions: the "possessing" class, and the ruling class proper. The "possessing" class is the class that derives its wealth primarily from the private sector through "comprador" and small manufacturing and banking activities. Generally, while it has strong licit and illicit linkages to the state, it does not control directly its apparatus. Those who have direct political control of the state represent what I define as the ruling class proper. They use the state as a private means of enrichment to acquire prebendary gains. They extract resources not only from their exploitation of subordinate classes but also from informal and formal taxation of the "possessing" class. In fact, if it wants to continue to do business and/or move into new ventures, the "possessing" class, more often than not, has to "buy" "protection" from the ruling class. There is thus a conflictive relationship between the two factions of the dominant class. These very real tensions dissipate, however, when subordinate classes mobilize to challenge the existing distribution of power, privilege, and property. At this point, the dominant class becomes united in its opposition to any fundamental change from below.

    An excellent survey of the Haitian class (Paris: Editions Karthala, 1994) structure is: ETZER Charles, *Le Pouvoir Politique en Haiti de 1957 A Nos Jours*, 13–176; see also: Robert Malval, *L'Année de toutes les Duperies* (Port-au-Prince: Editions Regain, 1996), 97–102.

33. Arthur Mahon, "Haiti: Une Dictature Rampante," *Rouge*, 144, December 7, 2000. See also: Gilles Danroc, "Imbroglio, Precarites et Democratie," *Diffusion de l'Information sur l'Amerique Latine*, Dossier 2358, Mars 1–15 2000, 1–2.
34. *Chimères* and *Zinglendos* are the Creole words used to describe violent and intimidating gangs. *Zinglendos* represents the new type of criminals who have emerged in the post-Duvalier period. The *Zinglendos* tend to be organized in armed gangs bent on the making of easy money through robberies of all kinds and drug trafficking. Their methods are extremely brutal, varying from intimidation to murder. The danger is that the Zinglendos will increase their power by becoming the armed wings of different political parties and of "druglords."

Chimères, on the other hand, has been associated with groups of "Lumpen" connected to Aristide's *Fanmi Lavalas* party; their function is to menace the opposition into silence. A senior member of *Fanmi Lavalas* told me, however, that he first encountered the *Chimères* in 1997 when he was campaigning in *Cité Soleil*—the largest slum of Port-au-Prince. He was informed by *Lavalassian* popular organizations that he could campaign in the *Cité* only if he was prepared to negotiate with the *Chimères*. Initially then, the *Chimères* seem to have enjoyed some autonomy from *Lavalas*. It is clear, nonetheless, that *Lavalas* had a privileged negotiating position with them. The *Chimères* would simply oppose the entrance of anti-Aristide forces into what they considered their "territory." The *Chimères* are thus a political entity associated with *Lavalas; Zinglendos*, on the other hand seem to be criminal elements linked to the drug trade and the old Duvalierist repressive security apparatus. Apparently, however, the distinction between the two groups is gradually vanishing. See: Jean-Michel Caroit, "En Toute Impunité les Chimères Font Régner La Terreur en Haiti," *Le Monde*, April 11, 2000, 4.

35. Bob Shacochis, "There Must Be a God In Haiti," in Edward Abbey, ed., *The Best of Outside: The First 20 Years* (New York: Vintage Departures, 1998), 306.
36. See Brian Weinstein and Aaron Segal, *Haiti: The Failure of Politics* (New York: Praeger, 1992).
37. *World Development Report, 1999–2000. Entering the 21st Century* (Oxford: Oxford University Press, 2000), 232, 236.
38. Mats Lundahl, *Peasants and Poverty: A Study of Haiti* (London: Croom Helm, 1979).
39. David Nicholls, *From Dessalines to Duvalier: Race, Colour and National Independence in Haiti* (Cambridge: Cambridge University Press, 1979); see also Micheline Labelle, *Idéologie de Couleur et Classes Sociales en Haïti* (Montréal: Les Presses de l'Université de Montréal, 1978).
40. Robert Putnam, *Making Democracy Work* (Princeton: Princeton University Press, 1993), 15–88.
41. T.H. Marshall, *Class, Citizenship and Social Development* (Westport: Greenwood Press, 1973), 71–72.
42. Ibid.
43. See: *Haiti en Marche*, "Carnaval Grands Mangeurs," Février 12–18, 1997, vol. XI, no. 1, 1–8; *Haiti en Marche,* Février 19–25, 1997, vol. XI, no. 2, 12.
44. Amy Wilentz, *The Rainy Season* (New York: Touchstone, 1989), 128, 224.
45. Ibid., p. 128.
46. Pierre Bourdieu, *In Other Words. Essays Towards A Reflexive Sociology* (Stanford: Stanford University Press, 1990), 9, 61, and 77.

Gérard Barthélémy uses "reflexes" to describe the same phenomenon in his work "Le Discours Duvalieriste Apres les Duvalier," in Gérard Barthélémy and Christian Girault, eds., *La République Haïtienne* (Paris: Karthala, 1993), 180:

First, there is a certain number of phenomena, habits, and beliefs that belong to the domain of general political behavior since 1804. These strong tendencies of national political life are not the prerogatives of any education in particular. They are deep behaviors belonging to a particular origin and history of which a long-term evolution is in progress, although at a very slow rate: that is why we refer to them with the term "reflexes."

Translated from the original French quotation:

Il y a d'abord un certain nombre de phénomènes, d'habitudes et de croyances qui appartiennent au comportement politique général, depuis 1804. Ces tendances fortes de la vie politique nationale ne sont l'apanage d'aucune formation en particulier. Il s'agit de comportements profonds, propres à une origine et à une histoire particulières dont l'évolution, à long terme, est en marche bien que de façon très lente: c'est pourquoi nous les désignons sous le terme de "réflexes."

47. Amy Wilentz in her book, *The Rainy Season*, reports the following comments of Aristide:

One good thing about all the assassination attempts is that I survived them. I know that sounds silly, or obvious, but think about it this way: Here you have a man, everyone is against him, the Macoutes, the hierarchy of the Church, the government, the Army. They go after him, with guns, machetes, stones. What happens? He survives. How do you think this makes people feel? I'll tell you. They think I'm protected. That I can't be hurt. That Jesus or the spirits are protecting me. That I am indestructible. This is great protection for me, because it makes a hired killer a little reluctant to take me on. Who wants to have on his hands the blood of someone the spirits protect? Worse, if he comes to kill me, the odds are, he thinks, that I will survive, and he will be punished. He thinks a powerful force is keeping me safe. (p. 234)

48. See, William Smarth, "Une Page de l'Eglise des Pauvres: Le Père Jean-Bertrand Aristide, Président d'Haiti," in Gérard Barthelemy and Christian Girault, eds., *La République Haitienne* (Paris: Karthala, 1993), 55–62.

49. Jean-Claude Jean and Marc Maesschalck, *Transition Politique en Haiti*, (Paris: L'Harmattan, 1999), 47, 48, 92. Translated from the original French quotation:

Même si "Lavalas" est un phénomène populaire, la "source" du courant réside dans une initiative personnelle fondée sur un leadership "charismatico-religieux." Malgré sa complexité et ses contradictions, tout le mouvement repose sur une seule personne, la seule capable de lui donner une impulsion et éventuellement de le relancer. Il y a donc une relation de dépendance complète à l'égard du leader. Or celui-ci a une pratique anti-organisationnelle déjà reconnue: aucun groupe n'a pu, dans son entourage, prendre un statut autonome et se développer librement. Spontanément, ce leader s'oppose a toute forme de contrôle de son pouvoir. Son action est stimulante pour les individus et peut produire des effets de masse, mais elle est destructurante pour un groupe qui cherche a se constituer et à s'organiser sur une base objective. . . .

Le rapport religeux au leader a plutôt favorisé l'attentisme, les relations fondées sur la confiance, a l'intérieur desquelles l'interpellation des dirigeants, l'autonomie des acteurs et la conscience critique, sont percues comme étant de la trahison. Le prêtre-président a neutralisé chez les masses le développement d'un jugement politique éclairé. Il a transformé un peuple revendicatif en une masse assagie, confiante dans la venue du miracle.

50. Meiksins Wood, *Democracy Against Capitalism*; see also her *Retreat From Class* (London: Verso, 1986).

51. Dietrich Rueschemeyer, Evelyne Huber Stephens, and John Stephens, *Capitalist Development and Democracy* (Chicago: University of Chicago Press, 1992).

52. World Bank, *World Development Report, 1997. The State in a Changing World* (Oxford: Oxford University Press, 1997), 220; World Bank, *World Development Report, 1999–2000. Entering the 21st Century* (Oxford: Oxford University Press, 2000), see also, Etzer Charles, *Le Pouvoir Politique en Haiti de 1957 A Nos Jours*, 44–47.

53. Haiti Solidarité Internationale, *Haiti Elections 1990. Quelle Démocratie?* (Port-au-Prince: Jean-Yves Urfie, 1990), 217–27.

54. Barrington Moore, Jr., *Social Origins of Dictatorship and Democracy* (Boston: Beacon Press, 1966), 418.

55. Ralph Miliband, *Marxism and Politics* (Oxford: Oxford University Press, 1977), 86–89.

56. Rueschemeyer, Huber Stephens, and Stephens, *Capitalist Development and Democracy*, 47.

57. Jean and Maesschalck, *Transition Politique en Haiti*, 84.

> Déstabilisés au plus profond de leur être par la misère, confrontés quotidiennement au désespoir de leur situation, beaucoup de militants ne sont pas en mesure d'investir dans la construction du long terme. Même s'ils en manifestent le désir, ils n'en ont ni la force ni les moyens. D'où la tendance à verser dans *l'activisme stérile*, à s'investir dans des actions aussi dangereuses que spectaculaires, don't l'impact dépasse rarement le court terme et l'éphémère.

58. Barrington Moore, Jr., *Social Origins of Dictatorship and Democracy*, see also Ralph Miliband, *Marxism and Politics*.

CHAPTER 14

PARTICIPATION, DEMOCRACY, AND HUMAN RIGHTS: AN APPROACH BASED ON THE DILEMMAS FACING LATIN AMERICA

José Thompson

This chapter analyzes the relationship between democracy and human rights, with an emphasis on the many faces of political participation. It then takes a look at recent conditions in Latin America in these areas in order to frame a modern-day meaning for postulates espoused or inspired by Thomas Jefferson.

THE CONNECTION BETWEEN DEMOCRACY AND HUMAN RIGHTS

Many attempts have been made to identify the ways in which respect for human rights is associated with a healthy democracy.[1] Nevertheless, relationships that have been described between the two concepts and their applicability, tend to be based mostly on assumption, without focusing on concrete cases taken from recent history.

The relationship between the two concepts—human rights and democracy—can be seen from various perspectives. For the purposes of this essay, we maintain that the use of one perspective over another in fact predetermines whether the resulting relationship between the two can serve as a useful tool for evaluating the strength, integrity and legitimacy of a democratic regime.

It should be understood that both institutions—democracy[2] and human rights, have evolved alongside the development of Western culture itself. Their common roots can be found in the splendor of Greek civilization, and both were built on the same foundations.

Greek democracy aspired to become a system of government that would provide an effective, legitimate means of making decisions for the entire citizenry.[3] By contrast, human rights doctrine revolves around certain central issues such as equity (Aristotle) and equality (the Stoic school) that arose from debates about justice as a value. The two schools find a common meeting ground in discussions of social justice.[4]

This relationship between the democratic form of government and the fundamental rights of the individual was later taken up with greater precision by the Classical School of Natural Law (Locke, Rousseau, and the movement of the Enlightenment). Thomas Jefferson drew inspiration from this school of thought and took it to new heights in his individual writings and in the Declaration of Independence of the United States, with the radical declaration that "[A]ll men are created equal, that they are endowed by their Creator with certain unalienable rights, that among these are Life, Liberty and the pursuit of Happiness . . ."

According to this view, which was distilled through many centuries and fertilized by many talents, the relationship between the fundamental rights of the human being and the valid exercise of power entails, or should entail, far more than just the means of electing rulers, and indeed should address their very legitimacy.

The position taken in this essay is that at present, there is a need for a formal frame of reference to govern the relationships between democracy and human rights, having the status of a full-fledged convention that would take its place in the constellation of international instruments already adopted to establish and protect human rights. Together, existing instruments reflect a consensus on how to define precisely the rights considered fundamental, and what implications they hold. This chapter focuses particularly on the provisions given in the American Convention on Human Rights (ACHR) and its Protocols.[5]

The texts of international human rights provisions uphold the existence of a working democratic regime as a requirement under the terms of the "political rights." Article 23 of the ACHR states,

1. Every citizen shall enjoy the following rights and opportunities:

    a. to take part in the conduct of public affairs, directly or through freely chosen representatives;
    b. to vote and to be elected in genuine periodic elections, which shall be by universal and equal suffrage and by secret ballot that guarantees the free expression of the will of the voters; and
    c. to have access, under general conditions of equality, to the public service of his country.

. . .

Clearly, essential characteristics of democracy as we know it today are fully embodied in this article: representation through the right to elect and be elected; universal, equal suffrage; equal opportunity to perform public service. None can claim therefore that human rights obligations have been respected in an atmosphere where the democratic system is not furthered and maintained.

"Political rights" are also classified as human rights related to freedom, and this holds serious implications for the application of guarantees in this area.[6]

The relationship does not stop there. Fundamental institutions of the democratic system are essential to ensure that human rights are respected, behaviors that violate these rights are punished, and the consequences of such violations are eventually redressed.

Clearly, the first institution involved is Justice.[7] It is the judge who is ideally positioned to receive complaints of human rights violations, investigate them, and if necessary, order corrective action or redress.[8] This means that without impartial, effective justice, there can be no real possibility for human rights to thrive.

The purpose of democracy is more than the simple exercise of a system of government, but also strives for citizen welfare. Human rights in their broadest sense provide a means to measure the quality of democracy with the use of universal parameters to which the countries themselves have agreed.[9]

We therefore maintain that human rights law exceeds the realm of provisions essential for the existence of democracy; and that democracy is the regime in which human rights can best flourish. Indeed, the yardstick of the health and quality of a given democracy can be found in the framework that human rights, broadly considered, can offer.

This useful relationship is revealed most clearly when applied to the area of political participation, a full range of activities that cannot be conducted without an effective human rights system. In the absence of such an environment, the activities that comprise political participation lose their true meaning, becoming a simple litany of rituals that could serve any cause, whether democratic or not.

THE ELEMENTS OF POLITICAL PARTICIPATION

Political participation has been defined in many ways. This paper will use the definition that the Inter-American Institute of Human Rights (IIHR) has adopted in its research and education activities. It is an approach that sees political participation as a complex notion best expressed in

> . . . all activities by members of a community that derive from their right to decide on their system of government, elect political representatives, be elected and hold positions of representation, participate in setting and preparing public rules and policies, and monitor the exercise of public duties entrusted to their representatives.

This concept embodies a number of elements that can be examined separately to facilitate understanding of the overall idea.

First of all, it is important to clarify that the term "community" is used advisedly and is intended to encompass multiple dimensions of a whole country or people. This means that participation in a broadly democratic society is exercised, not only in political processes at the national level, but also in similar fashion at regional or local levels. The term also allows for the great diversity of our world, recognizing that participation includes different ways of making political or shared decisions in indigenous, tribal or autonomous communities, even if their practices bear a strictly local stamp and cannot be extended to the national level.[10]

Within the broad concept of political participation, it is necessary to specify what is meant, first of all, by determination of a system of government. We acknowledge that democracy is the form of government required in the framework of human rights law, and we have stated that democracy is the only regime that can be considered legitimate in this context. The logical question is whether such a postulate indeed leaves any margin to exercise choice about a system of government. This question sparks debate about unity or diversity of concepts of democracy.

No longer is debate swirling around the question of whether socialist or "popular" systems were just as democratic as the Western or "representative" systems, as many claimed. This does not mean, however, that there are no "types" of democracy from which to choose. The range is broad, from the republican monarchy to the American-style presidential system. All communities are rightfully entitled to shape the characteristics of their own system of government, and by this very process, they enrich the growth of increasingly democratic solutions.

Recent discussions about the content and meaning of democracy in the Americas reveal that controversy continues to rage concerning the characteristics that need to exist in order to speak of true democracy. At one point, these discussions attempted to contrast the characteristics of "representative" from those of "participatory" democracy, as if the two could be separated.[11]

In today's world, it would appear surprising that anyone could question whether democracy is determined by its "representative" character. Not only is representation the defining quality of democracy, but even the framework established by human rights stipulates that electing and being elected, that is to say, the very practice of representation, are an essential component of political rights.

It would also appear, in view of conditions found today in Latin America, that representation alone is not enough to ensure democracy. Many citizens seek out new forms of direct participation in response to the aloofness of distant representatives who all too often prefer to act like "delegates." Thus, participation substantially enhances true democracy, but in no way contradicts "representation," understood correctly.

Many studies by the IIHR have added a third characteristic of democracy to broaden our understanding of this system of government as a rudder and an engine rather than a mere description of reality. This element is "inclusiveness," that is, the ability to recognize the full diversity of subjects, peoples, origins, and context, and reflect the variety and richness of ethnic groups, languages, perspectives, and cultures that comprise the mosaic we call the Americas.

This emphasis on "inclusiveness" finds expression in the movement to promote local government. In Latin America, the concentration of power in a central government has been, far and away, the dominant practice.[12] Large cities have become centers of development, with rural areas left behind. A comparison between numbers of inhabitants per local government in this part of the world and figures from France, the United States, or Switzerland very clearly illustrates the critical role of this factor in transforming government into something alien and distant.

A second element of political participation is the right to elect. Quite clearly, this right cannot be exercised fully unless a broad range of preexisting conditions and systems is in place. There is still no clear consensus on what conditions can be considered optimal to guarantee a secure vote, as an act of expressing citizen will. Nevertheless, a simple reading of the Convention text cited above clearly reveals that the right to elect cannot exist in the absence of a well-organized electoral process replete with guarantees.

The vote derives its very meaning from the presence of guarantees that elections are truly legitimate and the right to vote is universal, unhindered by considerations of gender or literacy.

In order for suffrage to become a truly universal right, at least the following conditions must exist: a reliable civil registry, up-to-date voter rolls, an identity document that confers the right to vote, an efficient electoral organization on voting day, and conditions that guarantee transparent and expeditious reporting of results.

A major concern today has been the problem of abstention. Should it be interpreted as a valid decision not to exercise the right to vote? Or instead is it a voluntary but dangerous move to elude the responsibilities of citizenship? Worse yet, it can be seen as a sign of mistrust in the democratic system itself and doubts over its very legitimacy.[13] The debate over the implications of this phenomenon is gaining strength and intensity. It is enough to look at percentages of electoral abstention in recent elections in Latin America.[14]

A third element in this useful concept of political participation is the right to be elected, closely associated with the right to elect. The derivations of this right and the conditions for exercising it could fill entire libraries, particularly when discussion touches on the sovereignty of the people and the nature of representation.

The right to be elected is affected by a longer list of restrictions than the right to suffrage;[15] moreover, it is exercised by means of an activity that depends on clear guarantees, and which in Latin America is subject to control by a variety of mechanisms. This part of the world has tended to understand the exercise of the right to be elected, not as an activity of representation, but as what some have called "delegation." The result is a type of despotism installed ritually every four, five, or six years when the population goes to the polls.[16] This view of the government ruler as one who clutches the reins of

power instead of one who leads as a representative or proxy has been reinforced by the spread of the presidential system of government.

At the other end of the spectrum of issues associated with the right to be
elected and serve as a representative is the whole matter of governance. This
refers to a representative's real capacity to make decisions when the centers
of power are so fragmented that the effective practice of government
becomes elusive.

The ideal exercise of the right to be elected must be found in the broad space
that lies between two undesirable extremes: ungovernability and despotism.

A fourth quality of political participation as posited here is the possibility
of influencing decisions on public rules and policies. It is a quality that gives
life to the adjective that this paper attaches to democracy: participatory. It
means that the citizenry can be consulted informally even outside the
framework of regular elections, or mechanisms can be created whereby organizations of civil society or individuals, without the intermediation of political
parties, can express their opinions, demand action, or propose initiatives.

This factor is important not only in preventing the tendency to exercise
"delegated power," but also to strengthen opportunities for negotiation and
for protection of minorities, contributing to the development of a democracy in action. The degree to which the mechanisms of direct or participatory democracy are universally exercised depends, clearly, on each country's
choices; but as they become widespread, life is more democratic at every
level—local, community, and citizen.

A final element of democracy is control over the exercise of public duties,
as embodied in the notion of accountability. This means not only that mechanisms exist to fight corruption and encourage transparency, but also and
more fundamentally, that those in public service are always willing to answer
for their actions in the jobs entrusted to them. It also refers to the real capacity
of society, whether through organizations or through individual actions by
each member, to oversee the performance of duties, the use of public
resources, and the fulfillment of commitments made during political
campaigns or when representatives are elected.

THE IMPACT OF HUMAN RIGHTS ON THE EXERCISE OF POLITICAL PARTICIPATION

This essay has stated and demonstrated that the relationship between democracy and human rights is revealed with special force through political participation. This is true not just because the essential rights that give life to
participation are protected in human rights instruments, but because participation would be impossible or meaningless in the absence of other human
rights. It is worthwhile to stop here and, for purposes of explanation, apply
this view to situations before, during, and after the moment when political
participation is most visibly expressed: elections.

In the first place, certain conditions are necessary for the healthy exercise
of political participation. These conditions must exist before elections occur
and synthesize the existence of respect for human rights, or lack thereof.

Without broad respect for freedom of expression,[17] the electorate has only limited capacity to hear and evaluate election promises and even to recognize the meaning of the electoral process and understand the impact of their own involvement in it.[18] This applies fully to freedom of the press, or the ability of the media to act freely and express opinions openly; but it is equally true in the realm upheld by European doctrines that place the right to information in a context much broader than simple expression of thoughts.[19]

The same can be said of rights such as freedom of association[20] and right of assembly;[21] if the latter were not respected, the possibility of translating party organization into a mechanism of direct communication with the population would be severely curtailed.[22]

The very act of holding an election depends on whether certain human rights do exist, including suffrage and the ability to aspire to an elected position. Although both can be analyzed in different ways, they do imply the existence and practice of many other rights.

Indeed, it is the principle of non-discrimination contained in human rights instruments[23] that gives meaning to a country's entire electoral system. Application of this principle results in elections that are universal and equality-based, which in turn lends them validity. Any restriction of the right to elect and be elected may be justified only with parameters that are nondiscriminatory.

Similarly, the existence of courts of electoral justice[24] is a response to the demands of human rights to establish institutional mechanisms for resolution of conflict.[25] The creation of such a jurisdiction needs to recognize applicable principles of due process that also appear in human rights instruments.

The exercise of political participation during the time following elections is also conditioned by or at least related to respect for specific human rights.

An understanding of the concept of a culture of accountability clearly reveals why it is so important to respect at least some margin for petition and access to public information. These are the tools that enable citizens and organizations to exercise effective supervision even under government leaders not characterized by transparency. If discovery of corruption does not culminate in effective action by the judicial system, an infraction of international human rights obligations has taken place.

These relationships are so obvious that they require little explanation. Even so, the simple fact of articulating their interconnections suggests much about the ways in which institutions of political participation should be applied and interpreted. Human rights doctrine, at its current stage of evolution, now covers more fields and entails more consequences for political participation.

Human dignity, a central concept in current trends of human rights thinking, calls for the effective exercise of a broad range of rights encompassing economic, social, and cultural conditions.

The 1993 Vienna Declaration that emerged from the World Conference on Human Rights was an attempt to close the gap between application of civil and political rights, on one hand, and economic, social, and cultural rights on the other. It proclaimed that all human rights are integral and interdependent.

Many experts continue to have reservations as to whether economic, social, and cultural rights can be classicized as human rights, a field they have customarily limited to "civil liberties." The truth is that a number of international instruments clearly enumerate the State's obligations concerning education, health, social security, employment, and working conditions. These include the International Covenant on Economic, Social and Cultural Rights, which applies both universally and regionally; the European Social Charter; and the San Salvador Protocol to the ACHR adopted by the countries of the Americas.

Economic, social, and cultural conditions can all be included as relevant considerations for evaluating the health of a political system and the extent of good government. The implications of including these factors may be massive, depending of course on the particular qualities of each one and how they are interpreted in any given society; but in all cases, they add new elements of analysis to the context of political participation.[26] The need to broaden the traditional view is most clearly visible when considering the many ways in which a State's investment in education has determined the population's ability to interpret electoral choices. The full spectrum of human rights provides an effective channel for such a study.[27]

The horizon for examining the development of political participation also widens if it includes collective rights such as those protected by Convention 169 of the International Labour Organisation, which targets indigenous populations. This approach considerably expands the universe of applicable rights and redefines the potential for political participation.[28] It also introduces such dilemmas as compatibility between community government systems and representation of indigenous leaders in national bodies.

To this can be added the broad interpretation that non-discrimination must begin by recognizing differences and translating them into an effective weapon against de facto inequalities. Some systems have addressed this issue by adopting quotas or numerical criteria.[29]

In such a context, the links between human rights and democracy take on a whole new profile. The concept of effective exercise of human rights assumes far greater dimensions when it is used as a tool to evaluate political participation in the framework of a system of government that seeks to transcend merely electoral democratic practice.

Naturally, an in-depth analysis of these issues would exceed the purpose and scope of this essay; the task at hand is to explore the extent to which this approach contributes to a better interpretation of current conditions in Latin America.

ACHIEVEMENTS AND DANGERS FOR DEMOCRACY IN LATIN AMERICA

For the first time in history, representative democracy has completely taken over Latin America, with the obvious and persistent exception of Cuba. Democratically elected regimes regularly yield power to similarly-chosen successors, albeit with certain up and downs. The armed forces, which still held so much political power only a few years ago, have returned to the barracks. Electoral structures and institutions have achieved great credibility.

The process is relatively recent. In the late 1970s and into the early 1980s, democracy was still an exception in this part of the world.[30] Only after 1985 did the wave of democracy begin to sweep away authoritarianism, usually consisting of autocratic military rule, and it quickly attained dimensions unimagined even by the greatest optimists. In only eight years, from 1992 to 2000, nearly 80 elections took place.

Significantly, very complex exercises such as the peace process in Central America explicitly seek agreements whereby countries commit to hold elections as means to gain the type of credibility so needed by most of the region's regimes.[31]

Democratic institutional structures as we understand them have been spreading steadily, with a few noteworthy exceptions; some fizzled, such as the attempted "self-coup" of President Serrano in Guatemala in 1993, and several that succeeded, such as the move by Peru's President Fujimori to dissolve Congress in 1992.[32] Fast-spreading reforms include separation of powers, checks and balances, and greater development of judicial and electoral institutions.

Specifically related to the central concern of this paper, channels for political participation expanded in the 1990s, and elections became a more intense, shared activity. The IIHR created its Center for Electoral Assistance and Promotion in 1983, and its activities began in 1984. By the end of 1985, the world's first association of electoral organizations had been created.[33] A rapidly-developing umbrella of associations is generating technical assistance projects based on the philosophy of horizontal south–south cooperation. The flaws and gaps in electoral mechanics and laws are rapidly being corrected in the countries of Latin America.[34]

The process of restoring democracy in Latin America has advanced parallel to the fight against human rights abuses. Authoritarian regimes thought they could gain legitimacy by claiming to defend national security from the onslaught of the communist threat.[35] The resulting dirty war was unleashed in this part of the world in the 1970s and much of the 1980s, with its wave of extra-judicial executions, disappearances, and trampling on fundamental rights and citizen freedoms. In both cases, success clearly depended on organization and action by the nongovernmental movement, which at its climax attained remarkable dimensions and acquired international presence and influence.[36]

To begin with, the nongovernmental movement smashed the silence that had shrouded the vicious violations of human rights, especially in countries of the Southern Cone. It reached out to international and intergovernmental institutions and civil society, denouncing violations over and over again, undermining the image being cultivated by authoritarianism and its methods. It learned from its research and by joining forces with others, and it created systems and networks so that its revelations would be more effective and so it could speak with greater force of the need to demand an end to the dirty war and replace the governing regimes.

Increasingly, the nongovernmental movement turned its energy to the fight to restore democracy, fanning demands for authoritarian regimes to depart. In its clamor to hasten the holding of legitimate, authentic elections, the movement joined forces with ever-broader and more diverse groups. As this fight began to gain ground, nongovernmental organizations were forced to take a new look at their agenda and reformulate their very mandate, faced with the disappearance of authoritarian regimes replaced by democratically elected governments.[37] Some nongovernmental organizations closed their doors, while others saw their field of action drastically narrowed. A few focused on demanding a reckoning from those responsible for the systematic human rights violations of the past. Others emerged with a focus on the recent democratization of Latin America.

This new group of NGOs has specialized in issues of democracy, ranging from electoral affairs to education for citizen participation. They have developed techniques, specialized know-how and critical skills that they are now sharing with their peers in Latin America or with emerging experiences in other latitudes, seizing hold of communication networks or forms of horizontal cooperation.[38]

Elections observation, an activity which flourished under the tide of emerging democracy,[39] has acquired a new dimension as entities of civil society develop and perfect techniques for observation in their own countries. One area of enormous complexity that NGOs have penetrated successfully is that of parallel or expeditious vote counts, especially valuable when doubts begin to swirl around the official process.[40] Other elements that stand out on the agenda of NGOs active in this field are the creation of new openings for discussion of political issues, and campaigns to encourage greater participation in critical stages of an electoral process.[41] Supervision of political campaigns, especially with respect to the use of the media, is a new field of action by civil society, together with the fight for transparency in the handling of public affairs.

Because of the work of NGOs and civil society in Latin America, the panorama today has changed substantially, regarding both democracy and the effective practice of human rights.

Of all the world's regions, it is Latin America that has most willingly submitted to international instruments for human rights protection. The ACHR has now been ratified by most of the countries and today has 25 member states. The Inter-American Court of Human Rights has jurisdiction over 22 countries around the Continent.[42] On September 11, 2001, a date now tragically remembered for other reasons, the Inter-American Democratic Charter was adopted, a milestone demonstrating the region's willingness to accede to international instruments. The Charter explicitly and irrefutably posits representative democracy as a condition for membership in the Organization of American States and the institutions for hemispheric integration, including the upcoming Free Trade Area of the Americas.

But this encouraging picture takes on dramatic nuances under close observation of the imperfections, inadequacy and cracks in Latin American democracy.

In the first place, growing disillusionment with democracy is increasingly in evidence. Public opinion polls consistently find lowest ratings for legislative bodies and political parties, without exception, and reveal a dangerous yearning for the easy solutions of authoritarian days.[43] The crisis of political parties has reached such an extreme in Venezuela, a country that was once home to an apparently healthy bipartisan culture, that it has catapulted into power a leader openly opposed to political parties, Hugo Chávez.

The current disillusionment with democracy feeds on two basic shortcomings of the system: its inability to respond to today's pressing, serious problems, and its failure to embrace all the diversity present in Latin America.

In the former case, the current structure has failed to provide answers in three critical problem areas: a shaky economic system, corruption, and lack of citizen security.

Argentina exemplifies dramatically the impact that economic problems can have on the health of democracy. Worse yet, one of the triggering factors for the crisis in that country was the application of measures recommended by international multilateral agencies.[44] In general, Latin America, which has the worst income distribution in the world,[45] has failed to inspire even moderate optimism for a more sound economic future, of the kind that should normally result from democracy and a legitimate climate of participation by the population.

A second problem, corruption, is striking a serious blow to the entire region, especially the political class, and discrediting the whole system. Part and parcel of the corruption problem is the sense of the impunity that arises when investigations and judicial processes rarely hold defendants accountable for problems proceeding from the upper reaches of political and economic spheres.

The third problem area is citizen insecurity in the face of spreading crime, international criminal activity, and ineffective security and judicial structures. This area is particularly vulnerable to extreme solutions that wrest legitimacy from Latin America's democratic regimes. Indeed, the population is crying out for quick fixes in the face of a spreading misperception that due process is responsible for the lack of effective actions to enhance security.

The second major factor eroding Latin American democracy is the exclusion of broad sectors of the population. Indigenous peoples and Afro-Latin Americans increasingly lack a sense of belonging, and women are faced with a patently unjust level of political representation.[46] A quick look at the membership of elected political bodies or the content of political or government agendas easily demonstrates that the lack of inclusion is echoed at every level. In these and so many other ways, the very idea of sovereignty of the people and for the people clearly comes into question.

As was found in the discussion of political participation, it is apparent here, too, that the dark side of Latin American democracy directly affects the exercise of human rights. Many issues are being addressed ineffectively, particularly those that entail economic, social, and cultural rights. These, along with demands for an independent, effective system of justice, in fact should hold top priority in government programs throughout the region, and all of them are associated with issues of corruption and insecurity. Exclusion itself clearly violates the derivative principles of equality and nondiscrimination, well established in the region and widely developed through individual instruments.[47]

If this relationship between the effective exercise of human rights and the health of democracy is real, it is time to ask how such an understanding can be useful in the political arena of our time, along with other expressions of political participation. It should also be asked whether these issues have any connection with what Thomas Jefferson left us, so long ago, as a philosophy and an aspiration.

HUMAN RIGHTS AS A GUIDE FOR GOOD GOVERNMENT AND A SEEDBED OF POLITICAL PARTICIPATION

The central thesis of this essay is that the effective exercise of human rights, the existence of a solid democracy, and a climate of healthy political participation are all intimately and inextricably bound together. This position has been clearly illustrated in the above sections, with special emphasis on the problems and dilemmas of Latin America.

In this text, we have expounded on the ethical duty of political practice to seek the welfare of all citizens, as was emphasized by Jefferson in his day and later by Locke and Rousseau; we have also seen that the explosion of theoretical development in the field of human rights has not been echoed by a parallel development of theories and proposals on dealing with the problems and challenges of our time.

Of course, it is one thing to describe the general process by which human rights have been formalized in legal documents. It is quite another to translate these formal theories into tools for political action, mechanisms for evaluating the health of democracy, or a yardstick to gauge the real possibility for exercising political participation.

The discussion does remind us, however, of the complexity of Jefferson's beliefs, the convictions of the great men of modern political thought, and others whose ideas laid the very foundations of democracy in all its forms. These great thinkers did not measure the legitimacy of government merely in terms of fair elections or clear separation among the branches of public power. Such factors, while important, are never sufficient. In the views of all these people, and in the United States Declaration of Independence, good government is a government that respects the rights of the population; it holds up these rights as the polestar of all its plans and actions, even going to the extreme belief that citizen obedience is no longer obligatory if these rights, natural and inalienable, are trampled upon by the powers of the day.

Today we have a more elaborate and precise corpus of human rights. We have a considerable body of international doctrine and jurisprudence with which to mark out our fields of action and catch a glimpse of what life would be like if these rights were fully respected. Still lacking, however, is a sincere effort to transform them into tools for the exercise of power in all the different settings where they should be applied.

In today's Latin America, this awareness should carry at least seven lessons for democracy:

1. Freedom is only one facet of the complex field of human rights; indeed, a democracy which fails to tackle problems of poverty and lack of economic opportunity is not fulfilling its promises.
2. Human beings are created equal; exclusions and discriminations, whether de facto or de jure, are contrary to human rights and delegitimize any democracy that tolerates them.
3. Citizen security is a top priority for the development of a society and a pressing demand of human rights; however, the fight against crime and violence must never resort to the very kinds of violations it seeks to combat.
4. Transparency and a culture of accountability are essential features of democracy and political participation. A climate of respect for human rights demands that all lawbreakers be brought to justice. Exceptions for the powerful only serve to undermine the solid foundations of democracy in the eyes of the people.
5. Participation is the lifeblood of true democracy. Society needs to take a stand by providing conditions for participation to be exercised through respect for fundamental human rights such as freedom of expression, association, and assembly, at least a minimum of economic and social conditions, and encouragement for decentralization of power through local governments.
6. The vote, the electoral system and strong institutions are indispensable for democracy; but they alone are not enough to ensure either political participation or the existence of a good government that never loses sight of its ultimate purpose: seeking the general welfare of those under its authority.
7. In today's world, the consolidation of a healthy, sound democracy and the effective exercise of human rights are matters that increasingly affect the entire world. The human face of globalization can be seen when people from every latitude take an interest and actively strive to support these causes.

This seventh lesson is particularly relevant in an essay intended to view current conditions under the light of the great ideas handed down to us by Thomas Jefferson. It reminds us that, in his day and in the face of challenges that lay before him, he drew on a great diversity of sources which lent much strength to the ideas that went into the United States Declaration of

Independence and Constitution. The doctrines of the great fighters for human rights served as Jefferson's inspiration and obsession. This thinker and statesman was not content merely to write down his convictions, but insisted on transforming them into real tools for struggle, change and government. This is perhaps the greatest lesson we can learn for the dilemmas of our day and for the anguish of Latin America.

NOTES

1. In this regard, see IIHR, *Estudios básicos de derechos humanos I* (San Jose, Inter-American Institute of Human Rights, 1994); Camargo, Pedro Pablo *Derechos humanos y democracia en América Latina: análisis comparativo,* (Bogota, 1996); Inter-American Commission on Human Rights, *Derechos humanos y democracia representativa* (Washington, 1965).

2. It would greatly exceed the scope of this essay to begin considering the abundant literature on the concept of democracy. As a start, see IIHR/Capel, *Diccionario electoral,* 2nd ed., vol. I (San Jose, Inter-American Institute of Human Rights, 2001), under the entry "democracy," 346 ff.

3. For more on the restrictive Greek concept of citizenry, see notes to www.lafacu.com/apuntes/politica/la_democ/; this is not necessarily comparable to the sense of universality apparent in discussions of the rights of the human being—with the possible exception of Plato's views on justice in the city, founded on differentiation and separation with harmony.

4. Notwithstanding disquisitions in the play *Antigone* on a law that overrides human law.

5. This choice was made not only because Latin America is the target of the essay, but also, as will be seen, because this Convention is the most explicit in associating democracy with human rights. Its text and emphasis can be compared with universal instruments such as the International Covenant on Civil and Political Rights and the Universal Declaration. Reference is also be made to instruments adopted under the International Labour Organisation.

6. Debates flourished for a long time concerning the categories of human rights and how choice of category would affect applicability; the emphasis for civil and political rights, the so-called freedoms, was on obligations for non-interference in the affairs of a State. See Cancado Trindade, Antonio, *El derecho internacional de los derechos humanos en el siglo XXI* (Editorial Jurídica de Chile, 2000), 59 ff.

7. The term as used here encompasses a number of expressions: "administration of justice," "judicial branch," and "State justice system"; at this point we will not linger over a discussion of the traits by which the various terms differ from one another.

8. The relationship between justice and human rights has been examined in a number of writings, including Thompson, José, *Acceso a la justicia y equidad. Estudio en siete países de América Latina,* Inter-American Development Bank/Inter-American Institute of Human Rights (San Jose Costa Rica, 2000), 416 ff.

9. The spread of the term "quality of democracy" in Latin America can be seen in *Diccionario electoral,* 109–21.

10. On these and related subjects, see inter alia Stavenhagen, Rodolfo et al., *Entre la ley y la costumbre: el derecho consuetudinario indígena en América Latina* (Mexico, 1990).

11. For example, an extensive debate that took place in the Organization of American States (OAS) in 2001 focused on the characteristics of democracy today. The purpose was to adopt the Inter-American Democratic Charter, the text of which can be consulted through the General Assembly of the Organization of American States (31st meeting, June 2001, San Jose). *Inter-American Democratic Charter*, http://www.oas.org.

12. See Brewer-Carias, Allen, *La opción entre democracia y autoritarismo*, inaugural lecture for the Fifteenth Conference of the Association of Electoral Bodies of Central America and the Caribbean (2001), 13–14: "Venezuela, with nearly one million square kilometers of land and some 24 million inhabitants, has only 338 municipalities. France, with half the size and only 59 million people, has 36,559 municipalities, that is, one hundred times as many . . ."

13. Some of the countries with the best-established democratic systems have the highest levels of abstention, as exemplified by the United States, where less than 50 percent of qualified voters cast a ballot. See Stephenson, D. Grier, *The principles of democratic elections*, U.S. Department of State, Democracy Papers, 5.

14. Presidential elections: Colombia (1998) first round 49 percent, second round 41 percent; El Salvador (1999) 61 percent; Guatemala (1999) first round 46.6 percent, second round 59.6 percent; Venezuela (2000) 43.69 percent.

15. Diverse causal factors enter into consideration, including age, ability and place of residence, that vary from one system to another.

16. O'Donnell, Guillermo, *Polyarchies and the (un)rule of law in Latin America*, (University of Notre Dame, 1999).

17. The ACHR states in Article 13. Freedom of Thought and Expression:

    1. Everyone has the right to freedom of thought and expression. This right includes freedom to seek, receive, and impart information and ideas of all kinds, regardless of frontiers, either orally, in writing, in print, in the form of art, or through any other medium of one's choice.

    2. The exercise of the right provided for in the foregoing paragraph shall not be subject to prior censorship but shall be subject to subsequent imposition of liability, which shall be expressly established by law to the extent necessary to ensure:
        a. respect for the rights or reputations of others; or
        b. the protection of national security, public order, or public health or morals.

    (. . .)

18. A unprecedented case occurred in Venezuela when, on May 25, 2000, the Constitutional Chamber of the Supreme Court, responding to a petition filed by an organization of civil society, ordered suspension of elections scheduled to take place just a few days later (May 28, 2000). Part of the reasoning was based on the inability of the electorate to be properly informed of the implications of the electoral process.

19. See Rebollo Vargas, R., *Aproximación a la jurisprudencia constitucional: libertad de expresión e información y sus límites penales* (Barcelona, 1992); and relevant provisions of the Spanish constitution (Article 20).

20. See ACHR Article 16.

21. See ACHR Article 15.

22. We subscribe to the notion that democracy as we now know it is fundamentally, although not exclusively, party democracy.

23. According to Article 1.1 of the ACHR:

> The States Parties to this Convention undertake to respect the rights and freedoms recognized herein and to ensure to all persons subject to their jurisdiction the free and full exercise of those rights and freedoms, without any discrimination for reasons of race, color, sex, language, religion, political or other opinion, national or social origin, economic status, birth, or any other social condition.

24. Various permutations of such systems can be found: some fit into the courts of the regular justice system, while others have a specialized, non-appealable jurisdiction; see the work of Orozco, Jesús, under the definition of the term "Justicia Electoral," *Diccionario electoral*, Vol. II, 752 ff.

25. ACHR Article 8:

> Every person has the right to a hearing, with due guarantees and within a reasonable time, by a competent, independent, and impartial tribunal, previously established by law, in the substantiation of any accusation of a criminal nature made against him or for the determination of his rights and obligations of a civil, labor, fiscal, or any other nature.

26. For an analysis of the implications of these rights, which derive primarily from the famous principle of freedom from want assumed by President Roosevelt in 1941, see Steiner, Henry and Alston, Philip, *International human rights in context* (Oxford, 2000), 237 ff.

27. The connection between certain economic rights and the full exercise of citizenship is nothing new, as recalled in Eide, Asbjorn:

> In 1950, T.H. Marshall focused on the historical development in the West of those attributes which were vital to effective "citizenship." He distinguished three stages in this evolution...Civil rights...the great achievement of the eighteenth century...political rights were the principal achievement of the nineteenth century...social rights were the contribution of the twentieth century, making it possible for all members of society to enjoy satisfactory conditions of life.
>
> In *Economic, social and cultural rights* (Marinus Nijhoff Publishers, 2001) 13; see also IIHR, *Los derechos económicos, sociales y culturales: un desafío impostergable* (San Jose, 1999).

28. Although available literature is very limited, those interested should consult: Oliart, Francisco, "Campesinado indígena y derecho electoral en América Latina," in *Cuadernos de Capel 6* (San Jose, 1986); *Nueva sociedad 153: Pueblos indígenas y democracia* (Caracas, 1998) and Guerrero, Andrés, "Poblaciones indígenas, ciudadanía y representación," in *Nueva Sociedad 150*, (Caracas, 1997).

29. For example, Argentine law dictates that party ballots should list women as candidates for at least 30 percent of all positions up for election, in proportions large enough to offer a realistic likelihood of being elected. No party ballots become official unless they fulfill these requirements. Specific quotas for women candidates have also been adopted in Paraguay, Bolivia, Costa Rica, Ecuador, Brazil, and the Dominican Republic. In Colombia, the constitution creates two special districts in the Senate for indigenous communities (Article 171) and grants legal facilities to create five special districts in the

House of Representatives for the ethnic groups consisting of blacks, political minorities, and Colombians abroad (Article 176).

30. In 1980, only Colombia, Costa Rica, and Venezuela were governed by explicitly recognized democracies.

31. This was the process that earned then-president Oscar Arias of Costa Rica the 1987 Nobel Peace Prize.

32. In a climate of very timid international response, Fujimori was able to establish a formally democratic regime that was increasingly corrupt and authoritarian until its stunning collapse in 2000.

33. Known as the "Tikal Protocol," its members were electoral institutions from Central America and the Caribbean. Capel still serves as Executive Secretariat for three different associations of electoral bodies in the Americas, including the United States and Canada.

34. To date, Capel has conducted nearly 60 such projects, whose impact was felt in 14 countries of Latin America. It has sent nearly 150 teams of elections observers, most of whose members have been representatives of electoral bodies from friendly countries; the intended purpose has usually been to provide an opportunity for exchange and cooperation in this area, rather than to evaluate the relative correctness of elections for consumption by local or international public opinion. Naturally, to have a more accurate idea of the true magnitude of work in this field, it would be important to consider the many similar institutions conducting activities of the same nature, including the International Foundation for Electoral Systems and the Unit for the Promotion of Democracy of the Organization of American States.

35. The classic text on the theory of "national security" and its impact on human rights is Montealegre, Hernán, *La seguridad del Estado y los derechos humanos* (Chile: Academia de Humanismo Cristiano, 1979); other sources include Hinkelammert, Franz J., *Democracia y totalitarismo* (San Jose: DEI, 1990), 273; Americas Watch, *With friends like these: the Americas Watch report on human rights US policy in Latin America* (New York: AW, 1985), 281; Barrientos, Lucrecia Elinor, *La seguridad del Estado y los derechos humanos*, (Guatemala: Universidad de San Carlos de Guatemala, 1992). 100.

36. It would greatly exceed the space available and the conceptual framework of this document to attempt an in-depth analysis of the nongovernmental movement in Latin America. For this fascinating subject, see Fruling, Hugo, ed., *Derechos humanos y democracia: la contribución de las organizaciones no gubernamentales* (Santiago: IIDH, 1991); Steiner, Henry J. et al., *Nongovernmental organizations in the human rights movement* (United States: Harvard University, 1991); Zalaquette, José, *The human rights issue and the human rights movement: characterization, evaluation, proposition* (Geneva: WCC, 1981), 65; Koojimans, P.H. et al., *The role of non-governmental organizations in the promotion and protection of human rights* (Leiden, Holland: Nederlands Juristen Comité voor de Mensenrechten, 1990). The Madres de la Plaza de Mayo and the Centro de Estudios Legales en Argentina, the Vicaría de la Solidaridad in Chile, the Servicio Paz y Justicia en Uruguay, Tutela Legal y Socorro Jurídico in El Salvador, to name only a few, clearly demonstrated with their struggles and their work the superhuman effort that was needed to bring down authoritarianism in Latin America.

37. On this process, which sparked what the IIHR called the "dilemmas and challenges" of nongovernmental organizations, see Ossa Henao, Carmela,

"Reflexiones sobre los desafíos de la protección de los derechos humanos desde la perspectiva de las organizaciones no gubernamentales" in *Estudios básicos de derechos humanos V* (Costa Rica, 1996); Cançado Trindade, Antônio Augusto, "Desafíos de la protección internacional de los derechos humanos al final del siglo XX," in *Memoria, seminario sobre derechos humanos,* (San Jose, IIHR, 1997), 97–124; Basombrío Iglesias, Carlos, ed., *¿Y ahora qué?: desafíos para el trabajo por los derechos humanos en América Latina* (Lima: Acción Ecuménica Sueca, 1996).

38. On these subjects, see IIDH/Capel *Boletín Electoral Latinoamericano,* number XVII (San Jose Costa Rica, 1997).

39. Observation of elections took on a range of international variables, such as the policy exemplified by the Carter Center in the private field and the Organization of American State in the intergovernmental sector, or electoral techniques practiced by elections agencies themselves in missions organized by CAPEL within the IIHR.

40. Particularly notable has been the work of the Peruvian organization Transparencia; see Vega, Rudecindo and Roncagliolo, Rafael, *Participación ciudadana y observación electoral* (Lima, 2000); Asociación Civil Transparencia, *Una historia que no debe repetirse: Perú, elecciones generales 2000: Informe de observación electoral* (Lima, Peru, 2001).

41. Because of its historical importance for the restoration of democracy in Chile, the civil association *Civitas* merits special mention. This group made contributions of overriding importance for the holding of the 1988 referendum which marked the downfall of the Pinochet regime. The organization's successor, *Participa*, is also one of Latin America's most active groups in electoral policy issues.

42. Up-to-date figures can be found at www.corteidh.or.cr

43. See IIHR/CAPEL, Rial, Juan et al. comp., *Urnas y desencanto político: elecciones y democracia en América Latina 1992–1996* (San Jose, Costa Rica, 1998); Cerdas, Rodolfo, *El desencanto democrático: crisis de partidos y transición democrática en Centro América y Panamá* (San Jose, 1993).

44. Four presidents in one month is a record difficult to beat, matched only by the external debt of nearly $132 billion.

45. See Inter-American Development Bank, *2000 Annual Report.* Washington, DC, IADB, 2000. At: http://www.iadb.org.

46. Levels of social exclusion in Latin America have inspired the new expression "development cum social exclusion." See CHALMERS et al., *The new politics of inequality in Latin America*, Oxford, 1997, 21.

47. The United Nations has produced a considerable body of normative provisions on this subject, perhaps broader than those of any other field of human rights. See the International Convention on the Elimination of All Forms of Racial Discrimination (1965) or the Convention on the Elimination of All Forms of Discrimination against Women (1979).

Bibliography

Bobbio, Norberto (1985). "La crisis de la democracia y la lección de los clásicos," In: *Crisis de la democracia*. Barcelona, Ariel.

Bobbio, Norberto (1985). *El futuro de la democracia*. Barcelona, Plaza y Janés.

Bobbio, Norberto (1992). *Estado, gobierno y sociedad: por una teoría general de la política*. México, DF: Fondo de Cultura Económica, 243.

Brewer-Carias (2001). Allan, *La opción entre democracia y autoritarismo...*, Inaugural address for the Fifteenth Conference of the Association of Electoral Bodies of Central America and the Caribbean. In press. Editorial: IIDI.

Camargo (1996). Pedro Pablo, *Derechos Humanos y democracia en América Latina: análisis comparativo.* Bogotá Grupo Editorial Leyer.

Cancado Trindade (2000). Antonio, *El Derecho Internacional de los derechos humanos en el siglo XXI.* Editorial Jurídica de Chile.

Cerdas, Rodolfo (1993). *El desencanto democrático: crisis de partidos y transición democrática en Centro América y Panamá.* San Jose, Red Editorial Iberoamericana.

Chalmers et al. (1997). *The new politics of inequality in Latin America.* Oxford, Oxford University Press.

Chomsky, Noam (1991). *Deterring democracy.* London: Verso Press.

Dahl, Robert A. (1972). *Polyarchy: participation and opposition.* New Haven: Yale University Press.

Dahl, Robert A. (1982). *Dilemmas of pluralist democracy: autonomy versus control.* New Haven: Yale University Press.

Dahl, Robert A. (1991). *Democracy and its critics.* New Haven: Yale University Press.

Eide, Asbjorn (1995). *Economic, social and cultural rights.* Netherlands, Martinus Nijhoff.

IIHR (1994). *Estudios Básicos de Derechos Humanos I.* San Jose Editorial: Inter-American Institute of Human Rights.

IIHR (1999). *Los derechos económicos, sociales y culturales: un desafío impostergable.* San Jose; Editorial Inter-American Institute of Human Rights.

IIHR/Capel (2000). *Diccionario electoral*, II ed. San Jose, Inter-American Institute of Human Rights.

IIHR/Capel, Rial, Juan et al. (1998). comp *Urnas y desencanto político: elecciones y democracia en América Latina 1992–1996.* San Jose, Costa Rica, Editorial: Inter-American Institute of Human Rights.

Inter-American Commission On Human Rights (1965). *Derechos humanos y democracia representativa,* Washington, Editorial: Inter-American Commission on Human Rights.

Inter-American Development Bank. (2001). *Annual report 2000.* Washington, DC: IADB.

Kelsen, Hans (1977). *Esencia y valor de la democracia.* Barcelona Ediciones Guadarrama.

O'Donnell, Guillermo (1999). *Polyarchies and the (un)rule of law in Latin America.* University of Notre Dame. Notre Dame, Indiana.

Oliart, Francisco (1986). "Campesinado indígena y derecho electoral en América Latina." *Cuadernos de Capel 6.* San Jose, Inter-American Institute of Human Rights.

Rebollo Vargas, R. (1992). *Aproximación a la jurisprudencia constitucional: libertad de expresión e información y sus límites penales.* Barcelona, PPU.

Stavenhagen, Rodolfo et al. (1990). *Entre la ley y la costumbre: el derecho consuetudinario indígena en América Latina.* México, Inter-American Institute of Human Rights.

Steiner, Henry and Alston, Philip (1996). *International Human Rights in context.* London: Oxford, Clarendon Press.

Stephenson, D. Grier (2001). *The principles of democratic elections.* U.S. Department of State Democracy Papers.

Thompson, José (2000). *Acceso a la Justicia y equidad. Estudio en siete países de América Latina.* Inter-American Development Bank/Inter-American Institute of Human Rights, San Jose, Costa Rica.

PART IV

RIGHTS, DEMOCRACY, AND POWER

CHAPTER 15

LOST ILLUSIONS

Stanley Hoffmann

Raymond Aron had two disciples who spent most of their lives studying international relations and took him, as Pierre Hassner has written, as "their master in international relations and in intellectual hygiene."[1] In these last years of my career, I want to tell Pierre how much I have owed him during the 47 years of our friendship and in particular how much his thoughts about the present state of the world have stimulated mine.

I

Hassner writes about violence. I prefer to use the word "war" in order to distinguish it from the "ordinary" forms of violence within states. In the twentieth century and in the beginning of the twenty-first, war has taken every imaginable form. Two world wars have ravaged the world, redrawn its map and given rise to the ideologies and regimes of fascism and communism. Multipolar international systems so scorned by Wilson and by American political science had insured a kind of peace between the fall of Napoleon and the crisis of summer 1914. No such system revived between the Versailles Treaty and Munich. The withdrawal of the new American superpower after 1918, the isolationism of the new Soviet Union and the distrust that Moscow provoked outside its borders left the game to tired European actors who haggled with each other, and to the Germans electrified by Hitler. At the end of World War II, atomic war made its atrocious and blinding appearance, the Cold War began quickly, and almost all international relations experts spent more than 40 years on the theory of a nuclear war between the two superpowers. Fortunately, we didn't get to practice it and there is a lesson here: the enormous importance of internal factors, that is, the liberal optimism of the Americans, persuaded that history had more chances than war in eroding communism, the Marxist-Leninist optimism, which promised the inevitable final triumph of communism over capitalism, the economic bankruptcy of the Soviet regime, and the success of an American economy that consumed Keynesian recipes despite the ideology of the free market. This period, which was so often nerve-wracking—at the time of the coup d'état in Prague, of the Korean War, of the Cuban crisis, of the missile crisis of 1983—appears today so far away that the American commentators, in this country, whose memory is short, now describe the Cold War as a confrontation between two conservative powers. As if a frenzied arms race had not wasted precious resources in a world that was largely miserable, and as if the need for nuclear prudence, the certain risk of nuclear retaliation in the case of a massive conventional confrontation, had not heavily weighed on daily life. Expelled from the realm of direct shocks between blocs armed to excess, war during those years shifted toward armed interventions without excessive risk: the United States in Central America and sometimes in the Middle East, the Soviets in order to maintain order in their empire and to intervene in Africa and Afghanistan. While the great powers discovered both the necessity of limits and of arms control, the huge movement of decolonization produced two eight-year–long wars (Indochina and Algeria) after having killed hundreds of thousands of people at the time of India's partition; it left the former Belgian Congo in chaos, and resulted in a cascade of wars between Arabs and Israelis.

After the collapse of the Soviet Union one was able to believe, during the period of a new war in the Middle East about Kuwait, that the model on which the United Nations had been built and that the Cold War had paralyzed— a world united against aggression under the leadership of the Big Five Security Council powers—was finally going to be realized. We know what happened instead. The alliance against Saddam Hussein left him in place, and we entered in a new phase about which Hassner and I have written a great deal, that of the failed states or pseudo-states, which were disintegrating, of ethnic conflict where people who had been obliged to coexist within the same borders rediscovered a hatred which many of them hadn't been aware of and manifested a passion for independence or revenge with extraordinary savagery. The nineties, in Yugoslavia, in central and east Africa, in many parts of Asia, were years of civil war (which Thucydides had already described in paragraphs saturated with emotion as particularly horrible) and so-called humanitarian interventions, or indeed calculated noninterventions as in Rwanda. The lessons are stark. The worst violence often occurs within legally sovereign states that happen to be fragile or artificial, and only force coming from the outside can reestablish an often fragile and artificial peace; but collective interventions occurring on behalf of humanitarian and non-egoistic interests meet strong opposition in a world where the national interest remains defined in narrowly self-interested terms. The responsibility the interveners inherit doesn't suit them very well and they usually try to escape from it as soon as the battle stops, or else by abstaining from action altogether.

At the beginning of the new century, in 2001, one could have thought that in the previous century, all conceivable kinds of atrocities, genocides, and collective crimes had been in evidence. We had continued to reason in terms of states, and of peoples aspiring to have their state, in other words, in the classical terms of international relations. The eleventh of September 2001 has revolutionized our perspective. We knew that modern weapons made security behind borders impossible to preserve. We now know that non-state actors—private groups or gangs—armed with weapons as unexpected as hijacked civilian aircraft, could settle almost anywhere and strike deliberately, civilians above all, and spread terror. Terrorism is not new, but it had mainly been an internal phenomenon except when the terrorists were serving a state that wanted to strike far from its borders. Now we can talk about a universal war that knows no borders, which makes the idea of victory perfectly unrealistic. From the New York towers to the discotheques of Bali, nothing is safe. The connection between state conflicts or ambitions and private terrorists deserves extensive study. A war against Iraq has provoked, even without any initiative from Saddam Hussein, terrorist movements of solidarity and new recruits for a Muslim fundamentalism very different from the lay skepticism of Saddam.

Thus, in less than a hundred years, our poor planet will have known a devaluation of borders and a multiplication of actors in the realm of violence. We have taken many steps toward the "one world" of the idealists but this single world has all the aspects of a jungle. The state, national or not, lives in a perfectly paradoxical situation. It is in fact open to all forms of insecurity coming from the outside, and its attempts to overcome these, by controlling access or by extensive surveillance of potential suspects, risk surrender it to various polices and to professional antiliberals, as well as creating citizens with limited rights and immigrants under suspicion, without however ever reaching the famous "homeland security" about which so much is said.

II

In the history of theories that try to give us concepts for understanding international relations and means of affecting them, there are two that matter most: realism and its variations, idealism and its variations. Realism guarantees a permanent "state of war," more or less moderate; idealism assures us that peace is possible. Both deserve the same bad grade. Neither one is of much help in the world as it is now.

The vast body of realism extends from Thucydides (who in any case is too subtle and deep to let himself be encased in a single "conceptual framework") to the present-day so-called neorealists (for whom a state's foreign policy is determined by the distribution of power in the world). It has, indeed, many merits. Interstate relations often resemble the realists' universe: the quest for power, state goals of security and domination, the preeminence of military power, profound differences between the big and the other actors, balancing and coalition exercises, and the like. These are indeed tunes. Hans Morgenthau devoted a hymn to them, Aron a symphony of Mahlerian complexity, Kenneth Waltz, the leading neorealist, a monody as simple as the music of Eric Satie. But the problems are numerous. The hypothesis according to which the combination of human nature and international anarchy obliges all the actors to pursue the same objectives— survival and security—is false insofar as states aim at more than survival and security, and make their own choices about everything else. Moreover, they also choose different kinds of security and different strategies of survival, depending on their geographical situation, their historical experiences, their means and their regimes. The most intelligent of the realists, such as the Swiss Arnold Wolfers and the Frenchman Raymond Aron, knew this extremely well. The years of humanitarian interventions have confirmed it. Furthermore, with some exceptions, the realists have underestimated the role of ideals (political or religious), of ideologies, and of purely or primarily internal factors in the making of foreign policy (Thucydides knew this well: the democracy of Alcibiades was not that of Pericles). The realists have overestimated hard power—military and economic—and underestimated what Joseph Nye (2002, 2003) is calling soft power,[2] the power to influence and persuade through one's culture and one's skills. Finally, just as realism has devoted too little attention to internal civil and political society, it has not paid any greater attention to transnational networks and flows, it has remained excessively skeptical of the effects of economic globalization, and has discounted what Aron called the germs of universal consciousness, for instance insofar as human rights are concerned. Realist thought is not false, but it is poor.

The problems of idealism do not always consist in a refusal to take into account the truths of realism: the description by Kant of the international "state of war" is very close to what Hobbes had said about it. But Kant's purpose is to leave this reality behind, and he has two powerful arguments. In his project on perpetual peace he counts on the moral sense of human beings, which he describes as still dormant; in his essay on universal history, he counts on the meaning and direction of history, which will lead to peace both through the intolerable exacerbation of violence and through trade and other benefits of enlightenment. Moral imperatives, in the purely normative version, and the cunning of nature, analyzed in the essay on history lead to the creation of representative regimes inside the states, and to interstate peace guaranteed by a sort of confederation, in international affairs. Kant inspired Woodrow Wilson and, more recently, John Rawls's (1999) philosophy.

In the eighteenth century, Rousseau had, in advance, rejected Kant's philosophy of history. War breeds only war, trade breeds only rivalries and conflicts, and peace can occur only in a world of very small states ruled by the principles of the social contract and concentrated on themselves—a world without the hostilities, envies, and temptations characteristic of interstate relations. And yet Kant was partly right. Trade, already celebrated by Montesquieu, is the motor of globalization today and it often softens or sweetens the relations among states; this is one of the factors that led to the Cold War's end just as it contributes today to the improvement of relations between the capitalist world and China. The possession by several countries of an "absolute" bomb has so far deterred its possessors from using it. However, representative democracy has not become the universal regime, and thus Rawls was obliged in his ideal theory to leave room for so-called "decent" but not democratic regimes. In the absence of such a universal domestic regime, representative democracies continue to make war against undemocratic regimes—we have seen this in 2001 and 2003—and interstate organizations, the modern version of Kant's confederation do not have the means to create peace in the world. Moreover, Kant and the idealists who, as good liberals, know the importance of civil society and of the "social capital" which animates it, leave much to be desired when the issues they have to face are the development of poor countries and the very complex effects of this global society, which they conceive simplistically as predominantly a good thing. Universal terror is an effect more visible than the peace, democracy, and individual emancipation that the champions of globalization expected from it. The call for a solidarity of human beings and states is not enough to remove, in a state or among states, the disruptions, inequalities, prejudices injustices, and exploitations that feed fanaticism and terrorist violence.

The most ambitious attempt at building an idealist philosophy of international relations at the end of the twentieth century and of his life was that of Rawls, but it is disappointing.[3] He leaves "non-ideal" philosophy to deal with many of the real problems of international relations, like secession, refugees, and migrations. The famous "difference principle," which within a "well-ordered" society, allows only those inequalities that aim at improving the fate of the least favored, is not, according to Rawls, applicable among states. In his conception of "the law of peoples," he does not define what constitutes a people, and he limits human rights to what "decent" societies are willing to accept. Finally, the principles that democratic states and decent states should agree on are strictly interstate norms: there is very little on the regulation of global civil society, we remain in the universe of the Westphalian treaties of the seventeenth century. This reminds us that the last word of liberal thought has always been the separation between a private, transnational society, which is largely commercial and financial, and an interstate society whose mode of regulation is cooperation among states, but not a real supranational cosmopolitan power.

III

When one thinks of the future of this fragmented as well as "globalized" world, one should remember the most disquieting of realists, Hobbes. He had the most somber conception of human nature and the most radical conception of the only way in which man can be saved from the annihilation that the war of all against all makes likely. The establishment of an absolute central power (although limited in scope) is the solution of the problem of insecurity in a community of human beings seeking survival. Hobbes considered that the international state of nature was less catastrophic than the state of nature within a group of men and women, and that one could therefore stay at the stage of competing states, but the reasons he was giving for this difference are no longer valid. His main concern was the survival of individuals. The state today is not capable any longer of softening, for its citizens, the effect of interstate wars, and of saving them from violent death. The potential of destruction of human lives both by the states' weapons and by terrorist forces has increased dramatically. The same logic that made the internal Leviathan necessary: the necessity to survive, should now push toward a worldwide Leviathan.

The least one can say is that it is unlikely. Individuals eager to survive, weak by nature, and relatively equal in their weakness, have an interest in signing a social contract in order to create a strong and protective power. States that are very unevenly powerful, and a heterogeneous mass of interests capable of ignoring borders, do not have the same reasons to precipitate themselves in to the arms of a world government. Many will believe, like terrorist groups, that they are perfectly capable of surviving in a world which is both open and anarchical. The strongest states prefer to insure their own security by traditional means, including alliances, even if these means are insufficient, rather than giving up their ambitions and transferring their powers to a world Leviathan, which risks oscillating between global tyranny and the kind of inefficiency of which the UN system gives a rather unappetizing idea.

In this decidedly unideal world, the substitute for a world Leviathan established by agreement is a global imperial power. Today it would be the American empire, which a large part of current American elites considers as the least bad of solutions, either, of course, from the viewpoint of the imperial power or, in Wilson's homeland, from the viewpoint of everybody's real interest all over in the world. But it is easy to see that this solution is not any more satisfactory. An empire "by invitation" and not by conquest, a world under the protection but not the direct control of the American "hyperpower" risks being constantly in a state of turbulence, like so many historical empires that were too large. It would be contested from within and rejected outside by all those who do not wish to submit. Moreover, the American empire lacks imperialists at home. The American people do not desire to be the international gendarme, do not have the enthusiasm or endurance necessary to ensure either human rights, or what is even better and more difficult, democracy elsewhere, nor the altruism indispensable to guarantee lasting and fair development, which would have to be less inegalitarian than the present one and therefore accompanied by much regulation rather than left to pure free-trade liberalism. Nor does it have the altruism necessary to preserve the cultural identity of the satellites and the clients without which they risk rebelling.

To paraphrase Aron: empire is unlikely, peace is impossible. We are at the bottom of an abyss (one can alas always go even more deeply and I cannot but think about a war against Iraq that would turn as hopeless as the Vietnam war, with external political and economic effects that would be much worse). Let me repeat: the state of war across borders affects both states and individuals or organized groups. Transnational society is made up in part of unrealized promises—concerning the environment, health, and human rights (except in the limited world of real democracy) and international justice—against which the single superpower is waging both war and blackmail. For the rest, this society often appears like a sum, not of benefits as Thomas Friedman seems to believe,[4] but of evils: in the economic realm, free trade and the freedom of financial flows, which are promoted by the hyperpower and by the international agencies it dominates, create as many woes among the weak as they create benefits for the strong; in the political realm the evils are called forced migrations, mafias, and terrorism.

How can we climb out of this abyss? There is no single recipe. There is much room for the joint efforts of international organizations (despite their meager means, their few powers, and their lack of autonomous military forces), of the states (despite the constant tension between long-term interests, which often argue for risky political interventions and for "sacrifices" to help the poor, and egoistic interests) and of nongovernmental organizations whose efficiency is often inferior to their goodwill. The ultimate objective should be a point of agreement between Hobbes and a renovated idealism: the gradual establishment of a world Leviathan that would be able to complement the efforts of the other actors and to deal with the problems those actors cannot solve either themselves or by the usual methods of cooperation. The unfortunately slow and shaky European Union could be an example if not a model. For humanitarian and security reasons, the grievances that are often at the root of terrorist actions should be seriously examined and dealt with. In a world in which democracy and hideous regimes, honest and corrupt ones, continue to coexist, the road to this global Leviathan will be neither easy nor swift but it is evident that it passes through a reinforcement, also gradual, of international organizations that need financial and military resources of their own. The current hostility of imperial America to the UN, its conviction that the "only superpower" can do without it and keep its hands free, makes such a reinforcement impossible. However, the interconnection of danger points in the world: the unwanted presence of foreign forces; the existence of tyrannical regimes often supported from the outside; the misery of millions of human beings; the degradation of the environment; the multiplicity of terrorisms; and the spread of weapons of mass destruction is such that it is more and more perilous to allow important conflicts to rot and to inject destructive passions into the hearts of people, as is the case of the Israelo-Palestinian conflict or that between India and Pakistan.

In a world full of axes and forces of evil, where most politicians think only of their own survival, it is more than ever necessary to remind oneself that the goddess Reason celebrated by the revolutionaries at the end of the century of Enlightenment has many weaknesses of her own. On the one hand, she will always find it difficult to rule a world of human beings, in other words, of emotions and passions, which often force reason to serve them, and that reason only occasionally succeeds in mastering. There are of course positive emotions and passions, but they rarely govern statesmen in international relations. The nongovernmental organizations that try to improve these relations usually lack the necessary means to repress the ethnic and ideological passions taking the Enemy or the Other as their target. As for religious passions, they have often fostered violence rather than the love that religion celebrates. Hence the precariousness of philosophies that put too much trust in the possibility and durability of fundamental agreement or consensus among allegedly rational human beings, such as the theories of Rawls and Habermas. On the other hand, one should not confuse instrumental and calculating reason, which can be deployed toward any kind of objective, including genocide, conquest of territory, or forcible collectivism, and Reason considered as the expression of a moral will as Kant had proposed in his *Metaphysical Foundations of Ethics* (a confusion between the two notions frequently afflicts the theory of rational choice). Realists who are pessimistic about human nature and about the nature of states but concerned about the survival, welfare, and freedoms or capabilities of human beings, idealists who are appalled by the huge gaps that separate ideal theory from reality, could meet around the non-metaphysical liberalism that Judith Shklar (1989) had called the liberalism of fear: a philosophy and ethics centered on the fight against and prevention of cruelty, oppression, fear, misery, and injustice, evils experiences by most human beings.[5] The realists will continue to think that it would engender only limited and temporary gains, the idealists would continue to be tormented by all the obstacles, but they should be capable of following that same road, even if they do not know how far they will be able to go.

Such a journey requires guides and leaders. The present American Administration, with its unilateralist instincts and lack of diplomatic skills, is, alas, not wise enough to guide, and too much addicted to bullying to serve as a wise leader. Let us hope that this is just a temporary divagation, and not one more lost illusion.

NOTES

This essay translated from the French, is a revised version of a contribution to a *Entre Kant et Kosovo*, a *Festschrift* for Pierre Hassner, edited by Anne-Marie Le Gloannec (Paris: Presses de Sciences Po, 2003).

1. *La violence et la paix* (Paris: Editions Esprit, 1995), 20.
2. See his *Paradox of American Power* (Oxford University Press, 2002), and *Soft Power*, Public Affairs, 2003.
3. See his *Law of Peoples* (Harvard University Press, 1999).
4. See his book, *The Lexus and the Olive Tree* (Farrar, Straus and Giroux, 2000).
5. See her essay, "The liberalism of fear," which first appeared in Nancy Rosenblum ed., Liberalism *and the Moral Life* (Harvard University Press, 1989).

CHAPTER 16

THE UNITED STATES, HUMAN RIGHTS, AND MORAL AUTONOMY IN THE POST–COLD WAR WORLD

Brantly Womack

The question of accountability in human rights can too easily be reduced to those of the measurement of behavior by a general standard and the execution of appropriate sanctions. The argument of this paper is that from the time of Thomas Jefferson to the present the application of standards is locked in a dialectic with the need for moral autonomy on the part of individuals and communities. The actor is not a transparent locus of behavior, and neither is the enforcer a neutral channel of universal justice. While the universality of human rights impels responsibility to general standards, the humanity of human rights requires attention to the concrete conditions of action.

Although the problem of moral autonomy does not reduce itself to questions of power, power provides the most fundamental context of action. This is as true for the enforcers of human rights accountability as it is for suspected violators. In situations of great disparity of capacity among nations, asymmetry puts the weaker side in a situation of vulnerability that heightens the sense of risk to moral autonomy, while the stronger side is not at risk and therefore can be insensitive to the implicit threat posed by intervention. The unique situation of the United States, the lone superpower in the post–Cold War era, therefore merits special attention.

In the new era the United States has shown tendencies toward unilateral universalism, and this is understandable given its short-term interests. But in fact, the world order is relatively stable despite asymmetry, and even the United States will need to be stable and credible if it is to achieve sustainable leadership. Meanwhile, intermediate powers, Russia, the major powers of

Asia, and Europe, have an immediate interest and special role in establishing acceptable patterns of multilateral human rights accountability.

THOMAS JEFFERSON, HUMAN RIGHTS, AND INDEPENDENCE

Thomas Jefferson's immortal words from the second paragraph of the Declaration of Independence are usually, and rightly, remembered as a statement of universal human rights. The rights of "life, liberty and the pursuit of happiness" are "inalienable" and "self-evident." Although they are "endowed by their creator," there is no theology suggested that might limit that endowment. "Their creator" is the least sectarian formulation possible. The reference adds cosmic resonance to the universal message, and the claim that "all men are created equal" has a righteous and moral momentum that would be lacking in "all men are equal."

Jefferson goes on to completely subordinate government to individual rights. First, it is the purpose of government "to secure these rights." But many governments might concede this as long as they remained the judges of their own performance. Second, however, governments derive "their just powers from the consent of the governed." Democracy was required for governmental legitimacy. Third, the legitimacy of democracy too is contingent on the people's opinion of its performance, for "whenever any form of government becomes destructive of these ends, it is the right of the people to alter or abolish it." In short, basic human rights are universal not only in the sense that they are not a product of time and place, but they are also prior to law and the state.

But this ringing endorsement of universal human rights is not used by Jefferson as the premise of an argument for the transparency of all public authorities and an untrammeled global accountability and responsibility for enforcement. Rather, it is part of a declaration of independence. The first paragraph of the Declaration makes clear that its primary purpose is to establish the "separate and equal station" of the United States of America. The point is not to achieve a redress of abuses, but to use the long train of abuses to justify a new political and moral space. Jefferson is not asking for intervention, he is asking for support from fellow Americans, sympathy from the English, and alliance from France and Spain. The rhetorical utility of inalienable human rights is to mobilize Americans, to cause division among the English, and to assure the French that the revolution is not a narrow or ephemeral movement but is grounded on abuses that all should agree are intolerable.

It is apparent from Jefferson's record of the debates surrounding the Declaration and from the changes in its text from his original draft that the major concerns of the Continental Congress were not the evidence for abuses, but about the more practical matters of uniting the states and securing aid from France.[1] Jefferson's condemnation of the slave trade was dropped because of Georgia and South Carolina's continuing imports and the involvement of northern states in slave transport. Rhetoric considered offensive to Englishmen was softened to avoid making unnecessary ene-mies.[2] A formal declaration of independence was necessary in order "to ren-der it consistent with European delicacy to treat with us, or even to receive an Ambassador from us."[3]

It was not assumed by the Continental Congress that a United States of America would be free from abuses. On the contrary, the discussion of the Articles of Confederation that proceeded immediately after the adoption of the Declaration of Independence showed considerable concern about the continuing condition of individuals and about the autonomy of states within the confederation. John Adams made the observation that "in some coun-tries the labouring poor were called freemen, in others they were called slaves; but that the difference as to the state was imaginary only."[4] In arguing for equality between large and small states in the confederation, Dr. Witherspoon argued that "If an equal vote be refused, the smaller states will become vassals to the larger, and all experience has shown that the vassals and subjects of free states are the most enslaved."[5] Although the abuses of the previous political order provided the justification for independence, it was not assumed that the new political order would be free from abuses, nor that it would permit the correction of abuses by external powers. Rather, it was assumed that the new political order would provide a more appropriate public space and structure for the American political community. A doctrine of universal, individual rights grounded a new claim for sovereign autonomy.

The Dialectic of Universal Standards and Moral Autonomy

The expression of universal human rights in a document arguing for sovereign autonomy might be taken as a concept before its time, like the notion of "all men are created equal" in a society that permitted slavery and the inequality of women. But the linkage between human rights and sovereign autonomy is, I would argue, neither an imperfection in the original articulation nor a contradiction. Rights and autonomy are in tension, but they are also essential to one another. There is a dialectic between them that requires constant, shift-ing, and unsatisfying compromises in concrete cases, and yet the absolutization of either, satisfying though it might be in terms of consistency, produces unfortunate and unintended results. The dialectic is further complicated by the existence of sovereign political units in an international context of unequal power and capacities.

The key to the dialectic is that while human rights reside ultimately in individuals, they are acknowledged and protected by communities. Rights are inherently individual because each person lives her or his own life. But the individuality of rights can be misleading. The term "rights" might seem to imply something that one either has or doesn't, and accountability might appear to be no more complicated in principle than looking for and correcting defective cases, much like quality control at a factory. But neither individuals, nor rights, nor systems of accountability exist in the abstract. Not only is life inextricably dependent upon others, beginning with one's parents, but the urge to protect another's welfare requires a fellow-feeling, an empathy, and that in turn implies a community. Thus while individual rights might be abstracted into identical and universal claims, the concrete reality of individuals and accountability for their welfare is inextricably bound with their communities. To reduce this thesis to an epigram, while the "rights" of human rights are universal, their humanity is communal.

The next step in exploring the dialectic of rights and moral autonomy is to examine the multiplicity, variety, and hierarchy of communities. Communities range from the intimacy of nuclear families to humanity in general. Clearly, any individual is simultaneously a member of a number of communities. One is usually a member of a family, a larger circle of relatives, a neighborhood, a locality, a state, and humanity, to name a few. If an individual looks up vertically at his or her communities, it might seem like an infinite and nebulous stack of memberships, some nesting inside others. However, if one looks horizontally from any of these levels, there are families to which this particular individual does not belong, localities other than this one, foreign states, and so forth. Only at the level of humanity is there one community, and that is the least tangible of the lot. Within even the most intimate communities the individual expects an acknowledgement of moral autonomy as well as of individual interests. Likewise, in a multiplicity of communities each community (whether families in a lineage organization or states in a multilateral body) expects not only that its interests will be part of the interests of the larger group, but that there will be appropriate space for its own discretion. If a larger community totally subordinates the deliberation and will of a smaller one to its own, then the smaller is no longer a community but has become at most merely an administrative subdivision of the larger. A hierarchy of communities creates a series of genus–species relationships in which each is at the same time in an inclusive relationship with the individuals and communities below it and relates as an individual to larger communities above it. "Above" and "below" do not suggest levels of power or authority, since a particular level of community can be strong and the level above it weak or even hypothetical. For reasons discussed below, in the modern world the state is usually the strongest level of community, while humanity often seems as inclusive as the sky but as far away.

The multiplicity of communities implies competition as well as hierarchy. A communal identity is defined not only by whom and what it claims for itself, but by what it regards as beyond its boundaries. The relationship with other, horizontal communities is not necessarily hostile, but it is necessarily an external relationship rather than an internal one. The two communities relate not as parts of a whole, but as self-regarding parties. Even if they are both members of a larger community and appeal to the common interest embodied in the encompassing community, to the extent that they have their own discretion and interests, they are in an implicitly competitive relationship with their fellows, because the identity of each excludes the other. In hostile competition a gain by the other is viewed as a loss. At its worst, a loss is not only a defeat, but also a weakening of ones own capacities for future struggles. Thus the anticipation of risk can sharpen competition into conflict. Just as individuals have a right to life, communities have an urge to collective survival. The *bellum aliquorum contra alios*, the war of some against others, can be as mortal as Hobbes's *bellum omnium contra omnes*, the war of all against all, and more destructive since the greater capacity of communities compared to individuals enables their struggles to be nastier, more brutal, and longer.

Of the various levels of community, states make the strongest claims to further the interests of their members, and their sovereignty defines the nodal point of political community. States assert exclusive control over territory and a monopoly over the means of violence within the territory, thereby assuming a privileged authoritative relationship, sovereignty, to the included individuals and communities. Clearly, as Jefferson indicates, the only reasonable justification for sovereignty is that it exists to secure the welfare of its members. On the other hand, to the extent that domestic competition requires regulation for the common good, the state's power is ultimately decisive in internal affairs. To the extent that external competition with other states must be addressed, the state not only has the obligation to defend the individual interests of its members, but also to preserve its own moral autonomy. It can bind itself through international agreements, but only if it merges into a larger sovereignty can it abdicate sovereign responsibility.

As a prerogative, sovereignty might be considered absolute in the abstract, just as individual rights might be.[6] However, the reality of sovereignty depends on the capacity of the state. Domestically, sovereignty is deficient to the extent that its order does not prevail. To the extent that laws and public authority do not matter and violence is not controlled, the state cannot function as the public guarantor. To the extent that the reach of the political community and public order only covers part of the population in its territory, the claim to sovereignty is incomplete. A limited state that guarantees zones of immunity to its citizens and their private organizations is quite different from a weak or incomplete state. A limited state is limited by constitutional consensus and effective in its areas of competence. A deficient public order creates a field of uncertainty, which some will experience as greater liberty, but many will experience as greater risk.

Externally there are also realities of power that shape the reality of sovereignty. In its international relations, the state must presume that it alone represents the common interests of its people, because it alone is at the apex of their political structure. However, while the statement that "all men are created equal" does not depart from the reality of the situation by more than an order of magnitude, the claim that "all states are created equal" seems strained. Not only are individuals more similar than states, they interact with a greater number of their fellows in a greater variety of situations. States are more like trees in a forest than like individuals in a marketplace: they are stuck in one place, their most important relationships are fixed by proximity, and size varies enormously. The external realities of power, therefore, vary from one relationship to the next and from one state to the next.

Any particular sovereignty is locked by its location into a matrix of regional relations in which relative capacities will remain fixed. Locational fates differ. Compare the bicoastal globalism of the United States with the landlocked situation of Laos. Relative capacities differ. Compared to Vietnam, the population of Laos is as small as Vietnam's population is when compared to China.[7] Given the range of capacities among states and their inability to rearrange and sort themselves according to size, a relationship between two roughly equal states is the exception. Asymmetric relationships are the rule.

Before moving on to the adrenaline-filled topic of asymmetry, it is good to reflect on the principles of accountability implied in the dialectic of individual rights and communities. They are complementary in the sense that each requires the other, and yet by that very logic of each requiring the other, the absolutization of either universal rights or community prerogative produces an irreconcilable conflict. There is a great intellectual and moral satisfaction in consistency and rationality. Therefore it is easy to become impatient with the tension and practical compromise implied in the dialectical relationship and to seek to tilt the game permanently (preferably some else's game) in favor of either individual or community. Just as an absolute theory of *Staatsräson* or of the "dictatorship of the proletariat" can warp a community's sense of its own moral limits, so too a self-righteous "white man's burden" to force other communities to abide by the standards of the enlightened can make the same problematic contribution to its object that religious crusades once made to the spread of religion.

Are there then any principles of accountability that could provide general guidance and yet encompass the dialectic of individual and community? It seems to me that a complementary pair of general principles could be derived from the initial Jeffersonian statement. First, that the advocacy of human rights should be premised on their self-evidence, and second, that there should be no intervention without representation.

The self-evidence of human rights is often taken as affirming that what appears to be common sense to me should be common sense to everyone. In other words, the content of human rights is assumed to be so unquestionable that any difference of opinion or action is based on willful distortion of the obvious. However, the notion of self-evidence can be transformed from being the bedrock of self-righteousness and preemptory superiority into a serious principle of respect for both rights and communities. If rights are self-evident, then the promotion of rights must be based on the common ground of humanity across communities. Like the concept of humanity in Confucian philosophy or the concept of reason in Plato's, the content of human rights should be able to be evoked by dialogue from any community. The trans-communal domain of human rights is one of discourse rather than one of force, because the common grounds of understanding and evidence are presumed to be present. If what is alleged to be a human right is in fact self-evident, it should eventually be persuasive.

The second principle, no intervention without representation, is founded on the notion that rights are endogenous to individuals in their communities, and therefore that the moral autonomy of the community must be respected in any action regarding its members undertaken by other communities. A community can bind itself to collective norms and institutions, such as the Charter of the United Nations. If it is not a party to the institution or group contemplating intervention, then the interveners are acting on the presumption that the community in question is either illegitimate or incompetent. Intervention against the will of the existing political community presumes that the individuals concerned are not being represented by their political unit, and therefore the principle of "no intervention without representation" is not being violated. Even if such a weighty presumption is justified, intervention carries with it the risk of creating a community that is warped by its external relationships.

If international interest in human rights should be founded on discourse rather than force and if intervention in states that are functioning as political communities should be based on representation, then there should be a presumption against unilateral sanctions such as trade sanctions in support of human rights causes. These may appear to add to the persuasiveness of discourse or to demonstrate sincerity, but in effect they are a poke in the ribs of the other political community. Compliance would be based on threat rather than on conviction, and the credibility of threat is based on relative power.

ASYMMETRY AND ACCOUNTABILITY

The relative capacities of states do not directly affect their competence as political communities; after all, larger states have more people to take care of, and their leaders can make grander mistakes. However, disparity in relative capacity creates a permanent situation of asymmetry in the opportunities and risks presented in international relations. To the extent that states are competitive, conflict and negotiation will occur not between equals, but between one confident of its strength and another aware of its vulnerability. Asymmetry in individual relationships is often counterbalanced by the general matrix of relations, since most states are larger than some but smaller than others.[8] However, at the smaller end of the spectrum there will be states that almost always negotiate from a position of inferior relative capacity, while at the other end the United States, especially in the post–Cold War era, is in a peculiarly dominant position in its international relations. Such differences affect not only bilateral relations, but also attitudes toward multilateral constraints.

Let us consider the pair of states, large A and small B. Even if we assume that the quantity of their interactions (trade and so forth) is equal, B presents proportionally a much smaller set of risks and opportunities to A than vice versa, and therefore normally occupies a smaller share of A's attention. By contrast, B pays close attention to A because of its vulnerability—perhaps too close attention, since A is not behaving toward B with a comparable intensity of interest. The possibilities for misperception that exist in any relationship are multiplied by asymmetry. A will tend to be sporadic in its attention to B, and impatience might induce it to push B with its superior power. B might well become paranoiac because of its vulnerability and interpret A's behavior as coordinated malevolence. Such tendencies are less evident in long standing and peaceful asymmetric relationships, but they easily lead to crises in novel situations such as the dismemberment of Yugoslavia.

The most obvious and direct effect of asymmetry on accountability and moral autonomy is that in fact the moral autonomy of only one side is vulnerable, and therefore that side is likely to be sensitive to infringements on its sovereign prerogatives. With regard to intervention, A and B do not stand equally behind a veil of ignorance awaiting justice. Intervention requires a stronger hand entering a weaker territory, and each knows how it stands with regard to relative strength. A can infringe on B's moral autonomy, but not vice versa. The principle of reciprocity rings hollow. It is hardly surprising that B would have a "victim mentality," since, if there is going to be a victim, it is clear who it will be. While A looks for wrongs to be righted, B waits to be wronged.

An indirect but equally interesting consequence of asymmetry is that A will be less interested in multilateral agreements and commitments regarding accountability. Multilateral constraints appear to be equally binding on all parties, but in fact they press hardest on those states that would otherwise have the greatest real discretion. Only if the constraint and its enforcement are in fact in the hands of the strongest does a multilateral commitment appear to its advantage, and this is more likely in the case of an alliance focused on a particular shared goal than it is in the case of a multilateral institution with a broad mandate. American diffidence toward the UN in general and in particular toward such bodies as the World Court illustrates this point.

This rather stark picture of bilateral asymmetry is modulated in many instances by circumstances. When one or both of the parties involved are intermediate powers, their general pattern of international relationships is mixed, and it is less likely that their self-perception and diplomatic habits in a particular case will be driven solely by the bilateral disparity. For instance, Mexico deals with Guatemala as a strong state with a weaker one, but it knows well what it is like to have a strong state on its border. Likewise, many relationships exist within a matrix of regional relations that usually buffers and stabilizes the bilateral relations embedded within it. The exceptions—Iraq's invasion of Kuwait, for example—often prove the rule. To return to Mexico and Guatemala, both parties must consider the attitudes of the other Central American states and of the United States in determining their bilateral relationship. Lastly, international organizations such as the United Nations, the World Bank, and the World Trade Organization (WTO) lay down regimes of expected behaviors, arbitration, and official inquiry that set a context of general expectations for some bilateral differences. The two sides may be imbalanced and playing their own game, but they are doing so in a world gaming-hall with its rules, officials and invigilators. All of these factors reduce the effect of bilateral asymmetry in many instances to insignificance.

For the United States, however, asymmetry is a major factor shaping its bilateral relationships and, more generally, its global hegemony. With a massive military superiority vis-à-vis any combination of other states, the world's largest economy and the world's third-largest population, the United States is in a unique situation. It is in the A position in all of its bilateral relations, and therefore it is uniquely tempted to generalize the perspective of the more powerful into a worldview of the most powerful. Its own moral autonomy is invulnerable, and its interest in human rights in other countries seems to be a matter of its own discretion. This creates the possibility of what I would call "unilateral universalism," the pursuit of ones own interests and preferences as if they were universally valid and applicable.

Although the United States has not always been the most powerful country in the world, its size and history have combined to deny it the experience of the B position. Since 1815 it has not fought a war with a more powerful nation. It has been isolated from European affairs by the Atlantic, and it quickly became the dominant player in hemispheric affairs. Geographic good fortune is amplified by a wealth of natural resources and a self-confidence of political and moral superiority and exceptionalism. The world wars reduced American isolation but confirmed its sense of superiority. In the Cold War, the rivalry with the Soviet Union was not with an equal but rather with a nemesis, an "evil empire." The major hot wars of the Cold War, Korea and Vietnam, were fought with smaller powers, and the frustration of the war in Vietnam is still an indigestible contradiction in the American self-image.

Despite the continuity of the American world role into the post–Cold War era, the situation of being the world's only superpower is novel. There is not even the shadow of a nemesis for negative guidance, and the plan for national missile defense would remove the ultimate sanction of mutually assured destruction. Despite the absence of a major enemy, the United States spent 36 percent of the world military budget in 1999, and after September 11 the military budget increased further. September 11 proved that the United States was vulnerable to terrorist attacks, but the war against terrorism in Afghanistan is being waged as a war of righteous annihilation more reminiscent of campaigns against American Indians in the late nineteenth century than of Cold War hostilities. Although the United States remains interested in international cooperation and alliances, policy is set by domestic perceptions and priorities rather than negotiated with other states. As the embargo of Cuba, the landmine treaty negotiations, and the Kyoto Protocol on global warming have illustrated, there are no other states or combination of states that can constrain American international policy, and, given the central role of the "hyperpower,"[9] its policy sets the global agenda. This is the essence of unilateral universalism.

If we apply our earlier discussion of the effects of asymmetry on human rights and moral autonomy to the current situation of the United States, it is clear that any other political community would feel vulnerable to the power of the United States and thus tend to be sensitive about preserving its moral autonomy in a problematic situation. On the other side of the coin, the United

States is in a position where it can easily ignore external opinion on its handling of human rights (the most obvious cases being the death penalty in general and in various individual executions), while the global reach of its influence creates an opportunity for it to influence other states by means of official policy, NGOs, and world media attention. In many cases American influence is welcomed by third parties as a latter-day *mission civilisatrice*, and the failure of the United States government to intervene is decried as a failure of virtuous policy-making. However, from the point of view of the targets or potential targets of such attention, American intervention in human rights results from the relative weakness of their state rather than simply from their faults, and it expresses the arrogance of power rather than the sincerity of conscience.

Accountability American style can be resisted not because there is a defense in principle of the wrongs alleged, but because the official attention of the United States, with its hand gently tapping on an array of sanctions, is considered a threat to the autonomy of any other political community and a preemptory denial of its integrity and legitimacy. If one compares, for instance, the Chinese white paper on human rights[10] to the annual U.S. State Department report on Chinese human rights,[11] China presents its accomplishments in human rights and challenges the appropriateness of American official interest. China does not respond to the abuses alleged in the American report, and this makes its white paper unconvincing to an American audience. However, to respond to charges within the framework of an official position paper (rather than, for instance, within the framework of nonpublic diplomatic inquiries or a joint human rights commission) would first of all acknowledge the legitimacy of American unilateral interest and second require it to admit, dispute or defend whatever charges the American annual reports contained. The Chinese government would no more condone torture than the American government would, nor would it be more able to prevent it absolutely—or to prevent allegations—than the American government. But the State Department report is taken as expressing a self-righteous superiority that implies a violation of China's moral autonomy.[12] In the broader case of "Asian values" that was raised in the mid-1990s, the crux of the matter was not the actual differences in values between Asia and "the West" (mostly the United States), but the devaluing of Asia itself that was implied in official interference.

The reciprocal question would be that of the accountability of the United States itself in the post–Cold War era. Since the heady days of the establishment of the UN the United States has been reluctant to sign agreements that might restrict its domestic options or license international oversight of its activities. The most recent example is the establishment of an International Criminal Court that was formally proposed by the Rome Statute of 1998. The United States signed the Statute in December 2000, but has not yet ratified it, and is unlikely to do so in the foreseeable future.[13] Moreover, the United States does not recognize the compulsory jurisdiction of the International Court of Justice, although it is a recognized practice to have an American justice on the Court.[14] Besides balking at such institutional constraints, the war on terrorism has led to situations that caused international alarm at the possible abuse of rights by the United States, most prominently in the case of incarceration of prisoners from Afghanistan at Guantanamo Base in Cuba. The American response to such pressures is often to investigate its own actions, but never to submit to an external review.

To end this section with the starkest possible formulation of the problem, unilateral universalism is pseudo-universal. To the extent that its will is determined by a single political community, the American reach into other political communities expresses the asymmetry of its power rather than a commonality of interests. To the extent that it is concerned with human rights, the superpower will find it more convenient to emphasize the prescription of behavior that will bring other states more in line with its own idealized domestic standards than to participate in multilateral institutions. In the dialectic of universal content of human rights and moral autonomy, universal content is reduced to unilateral identification of transgressions, and moral autonomy is simply ignored. On the other side of the coin, the challenge to moral autonomy that asymmetry necessarily presents to the weaker party is amplified by the arbitrariness of the superpower's unilateral behavior. Even if compliance with a demand is both feasible and reasonable, a weaker state (and every other state is weaker) must consider whether it is compromising its moral autonomy by acceding to the demand. Asymmetry has never been more important in world affairs.

BEYOND UNILATERAL UNIVERSALISM

In 2002, world news has been dominated by the continuing warfare in Afghanistan, the "axis of evil," prospects of an American invasion of Iraq, American nuclear strategy, and struggles between Israelis and Palestinians. All of these involve asymmetry, and they create a dark and swirling picture. Human rights have taken a back seat to more primal issues of security and the war on terrorism. One might shrink from extrapolating the future based on the first year of the millennium. Despite current anxieties, however, I would argue that unilateral universalism is just as likely to be an immature stage of a more stable world order as it is to be the beginning of its own destruction, and that intermediate powers and multilateral institutions might play a decisive role in the maturation process.

In contrast to the previous section, which concentrated on the differences created by asymmetry and the resulting structural misperceptions, the thesis of this section is that asymmetry is a normal condition. A preponderance of power does not necessarily create the ability to dominate others, and the long-term interests of even a superpower require broad cooperation. Intermediate powers have a special interest and a special role in stabilizing and institutionalizing asymmetry, thereby moving the current world order beyond its initial phase of unilateral universalism.

The disparity of capacity between A and B does indeed mean that A can do things to B that B cannot do to A, but one should not leap too quickly to the conclusion that the strong dominate and the weak submit. If asymmetry equaled control, then Vietnam would have been defeated first by the French, then by the Americans, and finally by the Chinese. Of course Vietnam was in no position to invade its opponents, and it won by counterposing the mortal concerns of an outraged community to the limited interests of its more powerful opponents. Nevertheless, Vietnam's fundamental contribution to contemporary world history is that it demonstrated that the powerful can be exhausted and frustrated by the weak. Relative power is not absolute power.

If we consider the likely American invasion of Iraq, which is a smaller, weaker and more isolated state than Vietnam, and one ruled by a much less popular and solid regime, the result is not clear. If the United States does not secure the cooperation of states in the region and beyond, then not only will its logistics be hampered, but all the non-cooperators would rather see it fail than succeed. Even if Saddam Hussein is easily overthrown, the problems of policing Iraq and creating a new regime would be tedious and difficult. Moreover, a commitment of this sort virtually precludes a major commitment elsewhere. By showing its strength unilaterally, the United States could lose its capacity to threaten others. And the alternative of coalition-building involves negotiating common purposes.

If even the world's only superpower is wise to have second thoughts about eliminating its most disliked opponent, the structure of the world order, asymmetric as it is, is rather stable. It is not an order of equals, but it presents a matrix of regional and global relationships in which each relationship is usually unequal but cannot simply be forced by the stronger side. If the matrix is stable, then it is to the interests of all that it is not misunderstood. Weaker powers would not want to be unnecessarily anxious about their vulnerability, and stronger powers normally would not want to have weaker powers unduly alarmed about their intentions, because alarm would lead to greater resistance rather than to greater compliance. Therefore, not only is there a normalcy to the worldwide web of asymmetric relations, but all have an interest in the proper understanding and management of these relations.

Even if both sides would prefer normalcy, the management of asymmetric relations is not easy. Because the relationship between A and B is not reciprocal, interaction has a different existential reality for each side. If both want to control misunderstanding and prevent unnecessary crises two techniques are available. First, specific issues and problems can be neutralized by being assigned to specialist commissions. If, for instance, a bilateral border commission is established to manage a common border, then the routinization of this arena for what otherwise could be hot disagreements lessens the likelihood of crisis. In the case of human rights, a bilateral commission is likely to be less incendiary and more effective than occasional confrontations or unilateral judgments of the behavior of the other. Second, a multilateral framework in which bilateral differences can be situated buffers the asymmetry of the bilateral relationship and deflects attention away from bilateral confrontation. For example, one of the basic functions of the WTO is to serve as a neutral venue for trade disputes. Clearly the UNHRC and other bodies serve this purpose in human rights. To the extent that bilateral relations can be neutralized by routinization and contained by mutual multilateral commitments, the tension inherent in asymmetric relations should be reduced.

Even though all sides have an interest in the proper management of asymmetry, the interests and possible contributions differ between the superpower and intermediate powers. For the United States, sustainable leadership requires a broad confidence that the order that it heads serves common purposes. For intermediate states, that is, for most states throughout the world but particularly for Russia, Europe, and the major Asian countries, there is a more immediate interest in predictable American behavior and effective multilateral institutions.

Clearly the United States is and will remain the central state of the current world order. If it slips from that position history's page will have turned, regardless of how the change occurred. It is in the strongest position of any Western power since the high point of the Roman empire. Yet it did not achieve this stature by conquest, but rather by the possession of a preponderance of national capacity and by leadership of like-minded states.

The fact of centrality and its advantageous position in all bilateral relationships creates a tremendous temptation for the United States to maximize its gains and to minimize binding precommitments. If the United States pursues its best possible bargain in transactions with other states, using issue linkage, its influence on third parties, and its strength in multilateral institutions, it may appear to be acting rationally by market standards. After all, the other side is pursuing its best possible outcome, and merely lacks the advantages of being the central player. However, such behavior in the long term sacrifices the relationship to the transaction. Because the center serves no interest greater than its own, it gives everyone else an incentive to reduce the power and discretion of the center. The center commands, but it does not lead. The rest comply, but they do not cooperate. A shift in relative capacity occurs because the center expends resources in sustaining compliance and the others sequester their resources or even combine against the center. Unilateral universalism is self-isolating and thereby undermines its own position.

By contrast, sustainable leadership builds loyalty and cooperation by maximizing the interests of the world order of which it is the apex rather than maximizing its own private interests as top dog. This does not require denying ones own interests. It does require self-control, however. The center would enter transactions with its handful of situational advantages tied behind it, and thus would be on an equal footing. The center would thus be credible as the representative of an impartial and inclusive order, and this would encourage loyalty and cooperation. The order, and the center's position, would be maintained by the quality of leadership and the cooperation that it induces rather than by the expenditure of resources.

With regard to human rights accountability, unilateral universalism preserves its own unaccountability and threatens the moral autonomy of the states that it targets. Sustainable leadership, by contrast, does not emphasize its own privileged position, and its efforts on behalf of human rights are cooperative and multilateral, thereby reducing the challenge to moral autonomy.

The prescription for sustainable leadership, like a recommendation for a healthy diet, is in principle neither impractical nor irrational nor self-denying. However, it requires the subordination of immediate gratification, and to a certain extent, of domestic political preferences, to long-term interests. It can operate only as an explicit ethical and strategic norm against the impulse to seize the apparent advantage. However, for intermediate powers, predictable and order-regarding behavior on the part of the central power is of both immediate and long-term interest. Indeed, the more that the central power acts like a hyperpower, the greater the general desire for sustainable leadership. In other words, precisely when the United States is least likely to control itself the rest of the world experiences the greatest desire to see it under control.

It might be said that Europe, the major powers of Asia, and Russia have a particular "civilizing mission" in the post–Cold War world, namely, to encourage the United States to move beyond unilateral universalism. As with all healthy regimens, the arguments most persuasive to the United States will be the effects of its own excesses rather than the advice of others. However, especially in the case of human rights accountability, the rational solutions are necessarily multilateral, and therefore initiatives can be taken by others.

The main advantage of the participation of intermediate powers in human rights accountability is that they are not the superpower. They have reason to be interested both in the moral autonomy of states as well as in the universality of human rights because they themselves are potentially at risk to unilateralism. The implicit challenge of their power to the moral autonomy of the state in question is less direct, and most of their mechanisms of accountability will be multilateral. Moreover, their own national experiences are varied, and thus cooperation requires movement away from a notion of self-evidence as "my common sense" to self-evidence as a common ground across national experiences. Discourse between former imperial powers like the United Kingdom and France and a former colony like India or a communist regime like China is likely to produce a notion of human rights that is less satisfying to reformers, but one that does not raise the confounding issue of external imposition. The war on terrorism will increasingly raise questions about human rights and the moral autonomy of states that will be more vivid to non-central actors because they are potentially at risk. The moral limits of the war on terrorism are likely to be less obvious to the United States than to other states.

What I am suggesting is not a confrontation between the United States and the intermediate powers and multilateral institutions, but rather their active and confident playing of a specific role in the matrix of asymmetric relationships. If the intermediate powers promote whenever possible the routinization of human rights issues in specialist commissions and the continued multilateralization of human rights institutions, they will be ahead of and possibly in tension with American short-term preferences, but in line with sustainable leadership. The recent loss and regaining of an American seat on the UNHRC is illustrative of the relationship. In 2001 the United States lost its seat on the Commission to Austria, due in part to inadequate preparation for the vote and in part to general alienation from American unilateralism. In 2002, the United States refused to run for a contested seat, and so Spain and Italy graciously abandoned their campaigns. Such deference is necessary to acknowledge the central position of the United States and to keep it invested in multilateral efforts. But the original defeat of the American candidate in 2001 was good as well, because it showed the autonomy of the Commission from American control. This is not only a useful lesson to the United States, it also strengthens the multilateral character of the Commission's actions.

In many respects the intermediate powers are already ahead of the United States. European foreign aid will soon be approximately four times that of the United States in terms of the percentage of GDP, and multilateral pressure is clearly involved in President Bush's recent initiatives in this area. The Kyoto Protocol on global warming and the Third UN Conference Against Racism are recent examples of massive efforts in which the United States played a less than leading role. Great Britain and Spain dealt with important issues of international liability for human rights abuses in the case of General Pinochet. China's efforts at domestic poverty reduction have been amazingly successful, and perhaps its methods could be applied more generally. Fortunately for the world, not everything that happens starts in Washington.

The point of this chapter is not for an American to tell others to do what they are doing already. Rather, its purpose is to suggest a theoretical framing of problems of human rights accountability at three levels. At the most basic level, rights are dialectically related to moral autonomy, and so the tension between the universality of human rights and national sovereignty must be managed, it cannot be eliminated. Community is the concrete humanity of human rights. Second, asymmetry creates a network of international relations in which the stronger and weaker parties are in radically different situations. The weaker side experiences more acutely the opportunities and risks of an asymmetric relationship, and may easily perceive it as more threatening than it is. The ultimate asymmetric relationship of the present era is that between the United States and everyone else, and it leads to the tendency toward American unilateral universalism.

Finally, despite asymmetry, the world order is more stable than it appears to be. Relative power does not equal absolute power, and the long-term interests of the superpower in sustainable leadership coincide with the short-term and long-term interests of everyone else in less arbitrary and more multilateral world politics. With regards to human rights in particular, the intermediate powers are in a particularly good position to formulate and advance initiatives because they are more sensitive to the problem of moral autonomy and less threatening in their activities, especially multilateral activities.

If Jefferson returned today, the appearance of the world would be novel to him, but its problems would seem familiar. When he declared certain rights to be self-evident, he knew that he was neither stating the obvious nor resolving the issue. Rather, he was setting a challenge with constantly shifting grounds and venues, one in which rights and independence would play roles that at times appear complementary, and at other times contradictory. He may have wished for the success that the American experiment has enjoyed, but he would not be surprised that challenges remain.

NOTES

1. See Thomas Jefferson, *Autobiography 1743–1790 With the Declaration of Independence*, in *Jefferson* (New York: Library of America, 1984), 13–24.
2. Ibid., 18.
3. Ibid., 16.
4. Ibid., 25.
5. Ibid., 29.
6. Of course, this paragraph cannot do justice to the complex realities of sovereignty, as the example of the European Union might illustrate. Nevertheless, I would argue (but not here) that most of the ambiguities could be located in conceptual space demarcated by scales of capacity and contractual limitation.
7. Roughly 1 : 16 in both cases. Of course, Laos also borders China.
8. The reader may well ask, "larger in what sense and by how much?" But I would like to postpone the distracting problem of measuring asymmetry. There are many cases where the disparity is so large and multi-dimensional that the question of measurement is peripheral, and one such case, the United States after the Cold War, is our ultimate focus here.
9. The term "hyperpower" was coined by French Foreign Minister Hubert Védrine (with Dominique Moisi) in *Les Cartes de la France a l'heure de la Mondialisation* [The assets of France in the era of globalization] (Paris: Fayard, 2000).
10. *Fifty Years of Progress in China's Human Rights* http://www.china.org.cn/e-white/3/index.htm.
11. U.S. Dept of State Human Rights Country Report on China 2000, at: http://www.state.gov/g/drl/rls/hrrpt/2000/.
12. In response, the Information Office of China's State Council issued a report, "Human Rights Record of the United States in 2001," on March 11, 2002. http://www.china.org.cn/english/2002/Mar/28587.htm
13. As of March 9, 2002, there were 55 ratifications and 139 signatories. Almost all European countries have already ratified. Sixty ratifications are necessary to establish the Court. See http://untreaty.un.org/ENGLISH/bible/englishinternetbible/partI/chapterXVIII/treaty10.asp#Notes.
14. In fact all of the five permanent members of the Security Council have been represented on the Court continuously, with the exception of China (from 1967 to 1984).

CHAPTER 17

HUMAN RIGHTS, PEACE, AND POWER

John Owen

INTRODUCTION

Thomas Jefferson was both a cosmopolitan and an American patriot. His assertions that "all men," that is, not only Englishmen or Americans, "are created equal" and "endowed by their Creator with certain inalienable rights," and his passionate support for the French Revolution, suggest that he felt linked on some political level with people around the world. Jefferson, however, was also an American patriot. He felt a special obligation to help the citizens of his *own* country enjoy their natural rights to life, liberty, and the pursuit of happiness.

In practice Jefferson reconciled his cosmopolitanism and his patriotism in the usual way: by assuming that if his own free country were secure and prosperous, liberty around the world would benefit. Robert W. Tucker and David C. Hendrickson title their book about President Jefferson's statecraft *Empire of Liberty*,[1] and that is precisely how the sage of Monticello thought of the United States. For him there was no contradiction in the expansion of political liberty and of U.S. power.

Like most statesmen, Jefferson fought too successfully against cognitive dissonance. In the real world, good things do not always go together. Expanding U.S. territory in the early nineteenth century meant treating with Napoleon Bonaparte, a dangerous despot (unrecognized as such by Jefferson), and also waging aggressive war upon Indian tribes. Jefferson turned out to be correct, however, that the territorial expansion of political liberty generally yields two political results: an increase in international peace, and an increase in the influence and power of the United States.

Human rights, that is to say, have political effects, by which I mean consequences for the distribution and uses of power within and among countries. By power I mean the ability to cause others to do what they otherwise would not. The political effects of human rights occur both within and among countries. Within a country, the institutionalization of human rights obviously alters the means of power and simultaneously devolves power away from the state and to the citizenry. If a person has certain inalienable rights—claims that the state in which he lives is bound to respect—then his state is by definition limited in the degree and ways in which it may coerce him. In other words, a state that upholds human rights is not only a more legitimate state in the moral sense. It also tends to be, in a descriptive sense, a less powerful, more domestically peaceful state.[2]

In this essay I argue that the diffusion of human rights recognition across countries yields analogous consequences in the international system. The growth in the number of states that uphold human rights has entailed an expanding zone of international peace and a rise in the influence and power of the United States. Human rights tend to be part of a package of institutions that we generally label liberal democracy. It is generally recognized that liberal democracies are far less likely to fight wars against one another than are other types of country.[3] Thus, as Jefferson's contemporary Immanuel Kant proposed, the club of liberal democracies—countries that uphold Jeffersonian human rights—is simultaneously a zone of relative peace.[4] At the same time, the liberal-democratic club tends to be a zone of American influence. It is not the case that all liberal democracies are thoroughly dominated by the United States, nor that they voluntarily follow America's lead in all matters. But they do tend to acquiesce to U.S. primacy more than do non-democratic states.

In the next section I argue that under current conditions human rights usually are part of a package of institutions generally termed liberal democratic. After that I consider the effects of the spread of liberal democracy (including human rights) on international peace. Finally I consider the effects of the same spread on American influence and power.

HUMAN RIGHTS AND LIBERAL DEMOCRACY

Human rights are not identical to democracy or popular rule. Historically, the notion that all persons had natural rights to certain goods, for example, to life or to fair judicial procedures, did not imply that they should be able to govern themselves. The English Bill of Rights of 1688, a reaction to the absolutist aspirations of the deposed James II, set certain limits upon the sovereign, but by no means brought democracy to England. Indeed, Thomas Hobbes, writing 37 years earlier, had used the notion of an individual right to life to justify an absolutist state: the Leviathan was set up to protect men from violent death.[5]

Nonetheless, as despots everywhere understand, human rights do by definition limit the power of the state vis-à-vis the subject or citizen. Even Hobbes allowed that subjects had a right to flee from capital punishment and hire others to serve in the military in their stead. Human rights thus devolve some measure of power from state to citizen, that is, they democratize a polity to some extent. Historically, the recognition of human rights seems to lead to democratization, however slowly or haltingly in many cases. In country after country over the past few centuries, subjects granted equal legal rights have come to think of themselves as equal in other ways to their political superiors, and come to demand more political power for themselves.

The relationship between human rights and democracy is so tight in today's world that we tend to see rights and self-rule as logically connected, parts of a general package of goods called liberal democracy. In a liberal democracy, the people—all adult citizens, excepting felons and the mentally incapable—may vote *and* are granted certain protections from the state. For their votes to have meaning, they must be uncoerced; for them to be uncoerced, they must be protected from state persecution. To us the England of 1688—legally egalitarian but politically hierarchical—seems anomalous, even nonsensical. Today, human rights practically imply or entail liberal democracy.

LIBERAL DEMOCRACY AND INTERNATIONAL PEACE

The consequences of liberal democracy, and hence of human rights, for international peace are profound. In recent years political scientists have come to a consensus that liberal democracies are significantly unlikely to fight wars against one another. Most agree that these countries fight non-democracies with at least normal frequency, and perhaps more often than chance alone can explain. But since the late eighteenth century liberal democracies have formed a zone of relative peace. No other system of government can claim to have formed such a zone: monarchies, communist and fascist countries have all fought one another with regular, grim frequency. Liberal democracies have conflicts of interest, and prior to the twentieth century often came close to war. But it is difficult to find a clear case of two liberal democracies at war. (A few partial exceptions exist, such as the Spanish–American War of 1898.) What is more, statistical studies show that the "liberal peace" cannot be explained by other factors such as alliances, geography, wealth, trade, or the balance of international power. (By no means do all political scientists accept the liberal peace thesis; as discussed below, the realist school is skeptical that liberal democracy, or indeed any domestic property of states, makes any difference in international politics.)[6]

It is worth noting in passing that many of Jefferson's cohort of late Enlightenment thinkers foresaw the liberal peace, although they would have termed it a "republican peace." Condorcet argued that peace would come with the spread of rational institutions and internationalist sentiment.[7] Paine averred that the European wars of the 1790s were caused not by French republicanism, as Edmund Burke and others argued, but rather by the resistance of the monarchies to the spread of political liberty.[8] Most famously, Kant predicted in 1795 that the world would move toward perpetual peace as republican government spread.[9] Jefferson never explicitly made such predictions, but they are entirely consistent with his thought on political liberty and statecraft.

Although a scholarly consensus exists *that* there is a liberal peace, no such consensus exists as to *why* there is a liberal peace. One hypothetical causal mechanism is liberal norms. People who believe in human rights and democracy, when involved in a dispute, tend to reject coercion in favor of discussion and compromise. These practices are normative not only within the state but also for diplomacy. In a dispute with a foreign country, liberals' default position is peaceful resolution. Just as lawless individuals within a state must be coerced, however, lawless states—those prone to use force over reason—must be coerced. Thus do liberal states sometimes find it necessary to wage war against non-democracies. (Some have noted analogies here to classical Islamic thinking about a dar al-Islam, or realm of submission to Allah, and a dar al-Harb, or realm of war.)

A second hypothesis is that liberal-democratic institutions constrain governments from going to war. War costs average people money by raising taxes and making consumer goods scarce; it costs many of them their lives as well. In a despotism, as Kant argued, the executive can decide on war without worrying about the costs borne by his subjects. In a liberal democracy, by contrast, the executive must take into account the possibility of being turned out of office in the event the war is unsuccessful or prolonged. Thus the executive of a liberal democracy will be more hesitant to go to war.

A third hypothesis is that liberals form a social group that, like all such groups, defines itself against a non-liberal "out-group." Liberal actors see themselves in a transnational struggle against illiberalism, a struggle that implicates the success of liberalism in their own countries. Thus they favor foreign policies that promote liberal institutions everywhere, and see their countries as being on the same side, as it were, as other liberal democracies. By contrast, they tend to be suspicious of non-democracies inasmuch as such countries are bound to oppose the spread of liberalism. These predispositions make it easier for liberal democracies to resolve disputes with one another peacefully, inasmuch as they trust one another more; they also make it harder for them to resolve disputes with non-democracies peacefully.[10]

A fourth hypothesis is that liberal democracies' relative transparency gives them advantages in cooperating. This hypothesis stipulates that war is always irrational *ex post facto*, since even victors lose resources; thus a satisfactory peaceful solution to any international dispute always exists *ex ante*. Wars happen, then, because countries misperceive one another's capabilities or intentions. Being transparent, liberal democracies are less able to hide their capabilities and intentions; for example, because opposition to war is openly expressed it is relatively easy to tell when a liberal democracy is bluffing about its willingness to go to war. Two liberal democracies in a dispute are thus less prone to misjudge each other, hence more likely to settle on a peaceful agreement, than a democracy and a non-democracy or, for that matter, two non-democracies.[11]

Each hypothesis is unsatisfactory in some way, and the true explanation may combine some elements of two or more of them. Whatever the explanation for liberal peace, the conclusion is clear. In today's world countries that uphold Jeffersonian human rights are virtually identified with liberal democracies. Liberal democracies are highly unlikely to fight wars against one another. It follows that the expansion of the number of countries that institutionalize human rights implies an expansion of a zone of peace. Peace is an intrinsic good. Countries at peace are also more likely to cooperate in other areas for mutual benefit. They are more likely to work together to secure such collective goods as economic and cultural openness, lower spending on armaments, ecological preservation, and international law. In a concrete sense, then, all countries that honor human rights have gained from the spread of human rights norms, because the number of countries with which each may more easily cooperate has correspondingly increased.

HUMAN RIGHTS AND AMERICAN INFLUENCE

But I suggest that one country has gained more from the spread of human rights norms than simply the gains from peace. The influence of the United States has grown along with the number of countries that honor human rights. Because influence enhances power—where power is the ability to get others to act as they otherwise would not—the power of the United States has grown along with the liberal-democratic zone. In other words, just as the recognition of human rights within a country shifts power from the state toward the citizens, the international spread of human rights norms shifts power from other countries—particularly non-democratic powers—to the United States. This contention is more controversial, and so I give it extended treatment.

Human rights and other liberal institutions act as carriers of American influence because they tend to yield leaders whose conception of their countries' interests overlap heavily with those of U.S. leaders. Non-American liberal leaders enact policies that solidify and deepen ties with the United States, including economic interdependence, diplomatic coordination, and military alliances. These policies in turn cause close relations with the United States to pay increasing returns over time, making it increasingly costly to loosen ties with America. By comparison, actors who reject human rights tend to see little or no overlap between the interests of their countries and those of the United States. Such actors resist U.S. influence and do what they can to reduce it in their countries and regions. Antiliberal actors are more likely to be in power in countries that do not honor human rights, such as China and Iran. It follows that the diffusion of human rights norms tends to bolster American influence and power.

By no means do all actors who favor human rights love all things American. Indeed, human rights advocates can be highly critical of U.S. foreign and domestic policies. Most European elites, for example, have come in recent years to see capital punishment, widely practiced in America, as a violation of human rights. They also have criticized the United States for unilateralism, particularly under the Bush administration. But these elites understand that, at bottom, the purpose to which America puts its international power and

influence is not a threat to their fundamental values. The United States is a
liberal hegemon, in contrast to past real and aspiring hegemons such as the
Soviet Union and Nazi Germany. Thus liberal elites find it irrational to
devote precious societal resources to overthrowing U.S. hegemony. At the
same time, the United States tends to treat fellow liberal countries more
kindly than non-liberal countries, thus reinforcing these tendencies.[12]

I shall present two types of evidence to support my contention that liberal-
democratic institutions tend to be carriers of U.S. influence. First, for at least
the past 450 years great powers have promoted particular institutions—be
they liberal-democratic or otherwise—in other countries as a way to spread
their influence. Second, today the liberal democracies of Western Europe and
Japan are not forming an anti-American alliance even though balance-of-
power politics would counsel them to do so.

Domestic Institutions and External Influence

The most venerable school of thought in international relations, realism,
downplays or even denies any significance for countries' domestic institutions. For realism, international relations is almost entirely about power.
States are envisaged as unitary actors responding in rational ways to shifts in
the distribution of power in the international system; their responses to these
shifts do not depend upon their internal institutions or indeed upon any
other domestic property. Thus if a great power wants to increase its influence
over another state, it may coerce the target state or bribe it, but it should not
bother changing the target's internal institutions; such efforts would only
waste precious resources.

If realism were correct about the irrelevance of domestic institutions,
however, then no country should ever promote particular institutions within
other countries as a way to increase its influence. A state might promote
domestic institutions to make itself feel virtuous or to appease some set of its
own citizens, but not to increase its power. A state should be especially
unlikely to use *force* to promote particular domestic institutions in other
countries when it is insecure, for example, when at war. In such times, realism says, a state must be most concerned not to waste its power. Nor should
we observe any tendency for countries that shift from one set of domestic
institutions to another—say, from Marxism-Leninism to liberal democracy—
to alter their foreign alignments.

In fact, however, evidence over the past several centuries strongly suggests
the opposite. In the 198 cases in which one state tried to impose domestic
institutions upon another in the past 450 years, the great majority took place
when the promoting country was involved in a hot or cold war. In other
words, states at war believe that they can increase their power by promoting
particular institutions in neighboring countries. What is more, the same
states impose domestic institutions repeatedly, suggesting that they find
the practice does indeed make them more powerful. In sixteenth- and
seventeenth-century Europe, Catholic states at war often promoted established

Catholicism in Protestant states, and Protestant states at war often returned the favor. In the eighteenth and nineteenth centuries, warring republics tried to impose republicanism abroad, warring monarchies did the same with monarchism, and Napoleon Bonaparte, almost perpetually at war, imposed his peculiar brand of bureaucratic authoritarianism. In the twentieth century, fascist, communist, liberal-democratic, and Islamist states forcibly promoted their institutions in other countries especially when they were at war or in danger of so being.[13]

In all of these periods, documentary evidence suggests that states do this so as to put into power in the target state people who are likely to align the target with them. For example, in 1558 England's Elizabeth I sent troops north to Scotland to help revolutionaries establish a Calvinist state. She did this not out of concern that Calvinism triumph—she was soon to persecute Calvinists within England—but rather because Scottish Calvinists were pro-English, while Scottish Catholics were pro-French. Elizabeth won her gamble: when the Calvinists did win in 1559, Scotland soon entered an alliance with England.[14] Two hundred and forty years later, the young French Republic imposed republican institutions upon the Low Countries, Switzerland, and Italy, precisely because republicans there tended to be pro-French and hostile toward monarchical Austria and Prussia. France thereby gained allies and buffers against the monarchical powers.[15]

Most familiar are the post–World War II U.S. impositions of liberal democracy on West Germany, Japan, and Italy, and the Soviet impositions of communism on most states of Eastern Europe. It is clear that each super-power was at least partially motivated by the desire to keep these target states in its sphere of influence. As a 1945 memorandum in the U.S. State Department, approved by President Truman, stated about Italy:

> Our objective is to strengthen Italy economically and politically so that truly democratic elements in the country can withstand the forces that threaten to sweep them into a new totalitarianism. Italian sympathies naturally and traditionally lie with the western democracies, and, with proper support from us, Italy would tend to become a factor for stability in Europe. The time is now ripe when we should initiate action to raise Italian morale, make a stable representative government possible, and permit Italy to become a responsible participant in international affairs.[16]

It is also clear that the strategy worked: no European democracy ever defected from the American bloc, and the only Soviet-bloc states that attempted to defect were those undergoing liberal reforms, namely Hungary in 1956 and Czechoslovakia in 1968.

Of course, during the Cold War the United States often failed to promote democracy and human rights in Latin America, Asia, and Africa, and indeed Washington sometimes actively opposed democracy in favor of anti-communist authoritarianism. Under some conditions, empowering the people in a country would weaken rather than increase U.S. influence, inasmuch as U.S. notions of social and political order are unpopular. For example, the paramount goal for most Third World elites during of the Cold War was national development, understood as independence from the old colonial West. Socialism was much more credible than capitalism as a strategy toward national development. Even those Third-World socialists who were democrats tended to tolerate communists, since they agreed on the need to build the state and to end the vestiges of imperialism. But because communists were likely to favor close relations with the Soviet Union, Washington could not tolerate the social democrats that tolerated them. The United States often threw its support behind authoritarian capitalists, who were tyrannical but anti-communist and hence likely to keep their countries out of the Soviet sphere of influence.[17]

What of the United States and the spread of human rights norms today? A brief look at the late 1980s and 1990s suggests that U.S. influence has increased along with the spread of liberal democracy, including the recognition of human rights. Under Mikhail Gorbachev's *glasnost* reforms, the Soviet Union began to recognize Jeffersonian human rights as never before. At the same time, Moscow ceased trying to counterbalance U.S. power at every move, to the point where it supported the U.S.-led coalition against its former ally Iraq in 1990–91. The Gulf War constituted a large if temporary increase in U.S. influence in the Middle East, and that increase could not have happened without Soviet cooperation. In the mind of the Soviet reformers, liberal reform implied cooperation with the democratic West. Furthermore, when the countries of the Warsaw Pact underwent their various liberal revolutions, they demanded an end to the Pact and hence to their alliance with the Soviet Union. During the 1990s most of these countries asked to join the U.S.-dominated North Atlantic alliance, and Poland, the Czech Republic, and Hungary have indeed joined. Here again, an extension of liberal democracy coincides with an extension of U.S. influence and power.

The major exception today is the Muslim world. Just as during the Cold War capitalism was unpopular among many Third-World elites, in the Muslim world the secularism that is part and parcel of political liberalism is violently rejected by many Muslim elites. Because of its vast power, Islamists see the United States as the engine of secularism. If despotic non-Islamist states such as Saudi Arabia, Egypt, and Pakistan became popularly governed countries under current conditions, the probability is high that they would become Islamist countries, and hence anti-liberal and (officially) anti-American. In other words, that Washington does not promote democracy and human rights in such countries does not contradict my argument; rather, it suggests that political liberalism itself has hardly any presence in those countries.[18]

The Persistence of American Primacy

One of the liveliest questions in international relations scholarship of late has been: why does the United States continue to enjoy being on the heavy end of an imbalance of international power, more than a decade after the collapse of the Soviet Union?[19] History and theory suggest that the international system abhors imbalances of power: some coalition of states ought to have formed to counterbalance American power. After all, in international politics concentrations of power are dangerous; all states should fear that the United States will use its power to impose its will upon them, and so a number of them should form an anti-American alliance. Indeed, in the 1990s a number of scholars (bravely but wrongly) predicted that within a very few years NATO would break up and the United States would face an alliance specifically designed to check its ability to affect world events.[20]

Some countries have moved toward counterbalancing U.S. power. Most strikingly, during the summer of 2001 Russia and China entered a treaty of cooperation that was explicitly directed at making the world "multipolar," that is, at eroding U.S. predominance. But this treaty was no alliance; neither country calls America its enemy; and one must ask why the Europeans, the Japanese, the Indians have not joined the Russians and Chinese. The closest the Europeans have come to counterbalancing America is the new 60,000-person Rapid Reaction Force (RRF), intended to respond quickly to humanitarian crises in the European region. The RRF cannot act without NATO, and hence U.S. approval. For its part, India has shown signs of shedding its historical anti-Americanism. (I leave aside the efforts of transnational terrorist networks to reduce American power; such networks are not states and so cannot by themselves counterbalance the United States. As seen below, however, it is highly significant that these terrorists are profoundly anti-liberal.)

I suggest that the primary reason why the United States faces no serious counterbalancing alliance is that political liberalism has penetrated potential challenger countries to a large degree. In Western Europe and Japan, liberal institutions are thoroughly ensconced, and the overwhelming majority of elites are political liberals—that is, persons who strongly favor institutions upholding human rights and democracy. Liberals certainly do not agree with the United States on every issue, and doubtless would prefer a world in which international power was more balanced. But at bottom, liberals do not worry about the purposes to which America is likely to put its power; or at least, they do not worry enough to favor devoting their own countries' precious resources to counterbalancing U.S. power. Military spending in Japan and Western Europe was flat or declining throughout the 1990s, even as U.S. predominance grew. At the same time, and for the same reasons, the United States tends to treat fellow liberal countries relatively well; for example, it never issues military threats against liberal states.

One way to understand the mechanism behind these cooperative relations is to recall the definition of war offered by Clausewitz: "an act of violence intended to compel our enemy to fulfill our will." To the extent that country A already fulfills the will of country B, the latter has no reason to attack the former. It is clear from U.S. promotion of human rights and democracy since the early 1990s that America's will for most other countries is that they be liberal democracies. Countries that are already liberal democracies, then, have much less to fear from American power.

Not so for non-democratic countries, or for illiberal actors residing in whatever country. Illiberal actors are those who reject human rights and democracy (perhaps for reasons that seem to them rational and even moral). Such actors fear the purpose to which the United States puts its power, and hence are more likely to want their countries to counterbalance U.S. power. (America makes exceptions where popular opinion is vehemently anti-American, as in the Middle East.)

Thus, among states with the potential to help mount a challenge to U.S. primacy, China is simultaneously the least liberal and has gone the farthest toward counterbalancing. Chinese military spending rose sharply in the 1990s. The two greatest jumps in that spending took place in 1992, following the Gulf War, and 2000, following the Kosovo air war. In both of those wars, American air power deeply impressed the Chinese military and communist party. It is clear from the writings of Chinese military officers that many consider the United States the main threat to Chinese security. At the same time, America tends to treat China with more suspicion and hostility than it treats liberal countries. For example, were China a liberal democracy, Washington would have little reason to preserve Taiwan's autonomy.

Russian policy toward U.S. primacy has been more ambivalent, and Russia is ideologically and institutionally torn between liberalism and authoritarianism. In the early 1990s Russia supported U.S. foreign policy almost sycophantically. Since then it has lurched back and forth between attempts to counterbalance U.S. power in Europe, Central Asia, and elsewhere, and standing aside as U.S. power expanded. Observers of Russian politics agree that liberals tend to favor a more pro-Western, pro-American policy, while communists and ultra-nationalists tend to be anti-American. At the same time, the United States treats Russia better than China—for example, by explicitly calling it a potential partner in missile defense—but less well than its European and Asian allies.

It is not the case that any advocate of human rights will think of himself as pro-American, or that liberal democracies will support all U.S. policies. It does appear to be the case, however, that liberals and liberal democracies are significantly less likely to devote resources to reduce or eliminate American influence. It follows that the diffusion of human rights norms carries with it, to some extent, an expansion of U.S. influence and hence power.

CONCLUSION

As Jefferson understood, the spread of political liberty entails shifts in political power. Within a country, the recognition of human rights brings civil peace and devolves power from state to citizenry. Among countries, the recognition of human rights, in conjunction with liberal democracy more generally, brings international peace and moves power toward the strongest liberal state, the United States.

I raise the international-political consequences of the diffusion of human rights norms not to argue for or against that diffusion. If one believes in human rights, one is bound to believe that the more countries recognize those rights, the better. After all, if *all* persons have equal human rights, then one cannot but celebrate the honoring of those rights. My purpose is rather to note that although human rights are an intrinsic good, they implicate the uses and distribution of power among nations. In particular, for policy-makers in the United States and elsewhere to ignore the increased influence that human rights recognition brings their country would be harmful to Americans and others. To wit, antiliberal actors in China, Russia, and the Islamic world fear human rights norms in part because they believe such norms are Trojan horses, bearing the American imperium. They are wrong— America is not an empire in the formal sense—but their fears contain a grain of truth: in today's world, whether one likes it or not, liberal democracy is an instrument not only of peace but of U.S. hegemony.

NOTES

1. Tucker and Hendrickson, *Empire of Liberty: The Statecraft of Thomas Jefferson* (New York: Oxford University Press, 1990).
2. For empirical support see R.J. Rummel, *Death by Government* (New Brunswick, NJ: Transaction, 1994).
3. For example, Michael W. Doyle, "Kant, Liberal Legacies, and Foreign Affairs, Part I," *Philosophy and Public Affairs* 12, no. 3 (Summer 1983), 205–35; Bruce M. Russett, *Grasping the Democratic Peace* (Princeton: Princeton University Press, 1992); John M. Owen, IV, *Liberal Peace, Liberal War: American Politics and International Security* (Ithaca, NY: Cornell University Press, 1997); Spencer Weart, *Never at War: Why Democracies Will Not Fight One Another* (New Haven: Yale University Press, 1998).
4. Immanuel Kant, "To Perpetual Peace: A Philosophical Sketch (1795)," in idem., *Perpetual Peace and Other Essays*, trans. Ted Humphrey (Indianapolis: Hackett Press, 1983).
5. Thomas Hobbes, *Leviathan* (1651), ed. C.B. Macpherson (London: Penguin, 1981).
6. Skeptical treatments of the liberal peace include Christopher Layne, "Kant or Cant: The Myth of the Liberal Peace," in Brown et al., *Debating the Democratic Peace*, 157–201; David Spiro, "The Insignificance of Liberal Peace," in Brown et al., eds., *Debating the Democratic Peace*, 202–38; Joanne Gowa, *Ballots or Bullets: The Elusive Democratic Peace* (Princeton: Princeton University Press, 1999).
7. J. Salwyn Schapiro, *Condorcet and the Rise of Liberalism* (New York: Harcourt, Brace & Co., 1934).
8. Michael Howard, *War and the Liberal Conscience* (New Brunswick: Rutgers University Press, 1978), 29.
9. Kant, "Perpetual Peace."
10. Margaret G. Hermann and Charles W. Kegley, Jr., "Rethinking Democracy and International Peace: Perspectives from Political Psychology," *International Studies Quarterly* 39, no. 4 (December 1995), 511–33; Weart, *Never at War*; Owen, *Liberal Peace*.

11. Kenneth A. Schultz, *Democracy and Coercive Diplomacy* (New York: Cambridge University Press, 2001); see also James D. Fearon, "Domestic Political Audiences and the Escalation of International Disputes," *American Political Science Review* 88, no. 3 (September 1994), 577–92.

12. For a fuller explanation, see John M. Owen, "Transnational Liberalism and U.S. Primacy," *International Security* 26, no. 3 (Winter 2001/2002), 117–52.

13. John M. Owen, "The Foreign Imposition of Domestic Institutions," *International Organization* 56, no. 2 (Spring 2002), 375–409.

14. R.B. Wernham, *Before the Armada: The Emergence of the English Nation 1485–1588* (London: Jonathan Cape, 1966), 244–57.

15. Cf. T.C.W. Blanning, *The French Revolutionary Wars 1787–1802* (New York: Arnold, 1996).

16. Cited in Owen, "Foreign Imposition of Domestic Institutions."

17. For more, see John M. Owen, "When Does America Support Authoritarianism?" paper presented at the annual meeting of the American Political Science Association, Boston, August 30, 2002.

18. Whether the United States and other countries can promote liberal democracy through more subtle means in the Muslim world is an open and vital question. It does appear important that secular democracy be shown to be a credible strategy for development in Muslim societies; hence the success of Turkey is probably crucial to the future of democracy in the Muslim world.

19. This section draws heavily upon Owen, "Transnational Liberalism."

20. Christopher Layne, "The Unipolar Illusion: Why New Great Powers Will Rise," *International Security* 17, no. 4 (Spring 1993), 5–49; Kenneth N. Waltz, "Structural Realism after the Cold War," *International Security* 25, no. 1 (Summer 2000), 5–41.

INDEX